Introduction to Responsible AI

Implement Ethical AI Using Python

Avinash Manure
Shaleen Bengani
Saravanan S

Apress®

Introduction to Responsible AI: Implement Ethical AI Using Python

Avinash Manure
Bangalore, Karnataka, India

Shaleen Bengani
Kolkata, West Bengal, India

Saravanan S
Chennai, Tamil Nadu, India

ISBN-13 (pbk): 978-1-4842-9981-4
https://doi.org/10.1007/978-1-4842-9982-1

ISBN-13 (electronic): 978-1-4842-9982-1

Managing Director, Apress Media LLC: Welmoed Spahr
Acquisitions Editor: Celestin Suresh John
Development Editor: Laura Berendson
Editorial Assistant: Gryffin Winkler
Copy Editor: April Rondeau

Cover designed by eStudioCalamar

Cover image by Fly D on Unsplash (www.unsplash.com)

Distributed to the book trade worldwide by Springer Science+Business Media New York, 1 New York Plaza, Suite 4600, New York, NY 10004-1562, USA. Phone 1-800-SPRINGER, fax (201) 348-4505, email orders-ny@springer-sbm.com, or visit www.springeronline.com. Apress Media, LLC is a California LLC and the sole member (owner) is Springer Science + Business Media Finance Inc (SSBM Finance Inc). SSBM Finance Inc is a **Delaware** corporation.

For information on translations, please e-mail booktranslations@springernature.com; for reprint, paperback, or audio rights, please e-mail bookpermissions@springernature.com.

Apress titles may be purchased in bulk for academic, corporate, or promotional use. eBook versions and licenses are also available for most titles. For more information, reference our Print and eBook Bulk Sales web page at http://www.apress.com/bulk-sales.

Any source code or other supplementary material referenced by the author in this book is available to readers on GitHub (github.com/apress). For more detailed information, please visit https://www.apress.com/gp/services/source-code.

Paper in this product is recyclable

Table of Contents

About the Authors

 **Avinash Manure** is a seasoned machine learning professional with more than ten years of experience in building, deploying, and maintaining state-of-the-art machine learning solutions across different industries. He has more than six years of experience in leading and mentoring high-performance teams in developing ML systems that cater to different business requirements. He is proficient in deploying complex machine learning and statistical modeling algorithms and techniques for identifying patterns and extracting valuable insights for key stakeholders and organizational leadership.

He is the author of *Learn Tensorflow 2.0* and *Introduction to Prescriptive AI,* both with Apress.

Avinash holds a bachelor's degree in electronics engineering from Mumbai University and earned his master's in business administration (marketing) from the University of Pune. He resides in Bangalore with his wife and child. He enjoys traveling to new places and reading motivational books.

Shaleen Bengani is a machine learning engineer with more than four years of experience in building, deploying, and managing cutting-edge machine learning solutions across varied industries. He has developed several frameworks and platforms that have significantly streamlined processes and improved efficiency of machine learning teams.

Bengani has authored the book *Operationalizing Machine Learning Pipelines* as well as multiple research papers in the deep learning space.

He holds a bachelor's degree in computer science and engineering from BITS Pilani, Dubai Campus, where he was awarded the Director's Medal for outstanding all-around performance. In his leisure time, he likes playing table tennis and reading.

Saravanan S is an AI engineer with more than six years of experience in data science and data engineering. He has developed robust data pipelines for developing and deploying advanced machine learning models, generating insightful reports, and ensuring cutting-edge solutions for diverse industries.

Saravanan earned a master's degree in statistics from Loyola College from Chennai. In his spare time, he likes traveling, reading books, and playing games.

About the Technical Reviewer

Akshay Kulkarni is an AI and machine learning evangelist and thought leader. He has consulted with several Fortune 500 and global enterprises to drive AI- and data science–led strategic transformations. He is a Google developer expert, author, and regular speaker at major AI and data science conferences (including Strata, O'Reilly AI Conf, and GIDS). He is a visiting faculty member for some of the top graduate institutes in India. In 2019, he was also featured as one of the top 40 under 40 data scientists in India. In his spare time, he enjoys reading, writing, coding, and building next-gen AI products.

CHAPTER 1

Introduction

In a world permeated by digital innovation, the rise of artificial intelligence (AI) stands as one of the most remarkable advancements of our era. AI, the simulated intelligence of machines capable of mimicking human cognitive processes, has ignited a transformative wave that spans across industries, from health care and finance to education and entertainment. As the boundaries of AI continue to expand, so too does its potential to reshape the very fabric of our society.

In this chapter, we shall embark on a journey to explore a concise overview of AI and the vast potential it holds. Subsequently, we will delve into the compelling reasons behind the significance of responsible AI. In the end, we will cast our gaze upon the foundational ethical principles that underpin the realm of responsible AI.

Brief Overview of AI and Its Potential

Artificial intelligence, once a realm of science fiction, has evolved into a transformative force shaping our contemporary world. This technological marvel, rooted in the emulation of human intelligence, has unveiled an era of unprecedented possibilities. In this section, we will delve into a succinct exploration of AI's foundational concepts, its diverse manifestations, and the remarkable potential it holds across various domains.

A. Manure et al., *Introduction to Responsible AI*,
https://doi.org/10.1007/978-1-4842-9982-1_1

Foundations of AI: From Concept to Reality

At its core, AI is an interdisciplinary domain that seeks to develop machines capable of executing tasks that typically require human intelligence. It encompasses a spectrum of technologies and techniques, each contributing to the advancement of AI's capabilities.

AI's journey traces back to the mid-twentieth century, with pioneers like Alan Turing laying the groundwork for the field's theoretical underpinnings. The development of early AI systems, often based on symbolic reasoning, marked a significant step forward. These systems aimed to replicate human thought processes through the manipulation of symbols and rules.

However, it was the advent of machine learning that revolutionized AI's trajectory. Machine learning empowers computers to acquire knowledge from data, allowing them to adjust and enhance their performance over time. Neural networks, inspired by how human brains work, enabled the emergence of revolutionary deep learning technology, responsible for groundbreaking achievements in vision (image recognition), speech (natural language processing), and more.

AI in Action: A Multifaceted Landscape

AI's potential is vast and extends across a spectrum of applications, each amplifying our ability to address complex challenges. One of AI's prominent manifestations is in the realm of data analysis. The ability of AI algorithms to sift through vast datasets and extract meaningful insights has revolutionized industries like finance, health care, and marketing. For instance, financial institutions employ AI-powered algorithms to detect fraudulent activities and predict market trends, enhancing decision making and risk management.

AI's prowess shines in its capacity for automation. Robotic process automation (RPA) streamlines routine tasks, freeing human resources for more strategic endeavors. Manufacturing, logistics, and customer service have all witnessed the efficiency and precision AI-driven automation can bring.

Another notable domain is natural language processing (NLP), which empowers machines to comprehend and generate human language. This technology finds applications in chatbots, language translation, and sentiment analysis, transforming the way businesses engage with customers and analyze textual data.

Health care, a sector perpetually seeking innovation, is experiencing a revolution through AI. Diagnostic tools fueled by AI aid in the early detection of diseases, while predictive analytics assist in identifying outbreaks and planning resource allocation. The amalgamation of AI with medical imaging is enhancing diagnostic accuracy, expediting treatment decisions, and potentially saving lives.

The Promise of AI: Unlocking Boundless Potential

The potential of AI extends beyond incremental advancements; it possesses the capacity to reshape industries, enhance our quality of life, and address societal challenges. One such promise lies in autonomous vehicles. AI-powered self-driving cars have the potential to reduce accidents, optimize traffic flow, and redefine urban mobility.

In the realm of environmental conservation, AI plays a pivotal role. Predictive models analyze complex climate data to anticipate natural disasters, aiding in disaster preparedness and response. Additionally, AI-driven precision agriculture optimizes crop yields, reduces resource wastage, and contributes to sustainable food production.

Education, too, stands to benefit immensely from AI. Personalized learning platforms leverage AI to adapt content to individual learning styles, ensuring effective knowledge absorption. Moreover, AI-powered tutoring systems provide students with immediate feedback, fostering a deeper understanding of subjects.

Navigating the AI Frontier

As we stand on the precipice of the AI revolution, the horizon brims with potential. From streamlining industries to revolutionizing health care and empowering education, AI's transformative influence is undeniable. Yet, with its soaring capabilities comes the responsibility of harnessing its potential ethically and responsibly, ensuring that progress is accompanied by compassion, inclusivity, and accountability. In the chapters that follow, we will delve deeper into the ethical considerations and guiding principles that underpin the responsible integration of AI into our lives.

Importance of Responsible AI

In the ever-evolving landscape of technology, AI emerges as a beacon of innovation, promising to revolutionize industries, elevate human capabilities, and redefine problem-solving paradigms. Yet, as AI takes center stage, the imperative of responsibility looms larger than ever before. In this exploration, we delve into the profound importance of responsible AI, unraveling its ethical dimensions, societal implications, and the critical role it plays in shaping a sustainable future.

Ethics in the Age of AI: The Call for Responsibility

As AI's capabilities flourish, its potential to influence human lives, societies, and economies becomes increasingly apparent. However, with this potential comes an inherent ethical dilemma—the power to create and wield machines capable of decision making, learning, and even autonomy. Responsible AI emerges as the lodestar guiding the development, deployment, and governance of AI technologies.

At its core, responsible AI calls for a deliberate alignment of technological innovation with societal values. It beckons developers, policymakers, and stakeholders to uphold ethical principles, accountability, and transparency throughout the AI lifecycle. Its significance transcends mere technology; it signifies a commitment to safeguarding human well being and ensuring equitable benefits for all.

Mitigating Bias and Discrimination: Pioneering Fairness and Equity

A glaring concern in the AI landscape is the potential for bias and discrimination to be embedded in algorithms. AI systems trained on biased data can perpetuate societal prejudices and exacerbate existing inequalities. Responsible AI takes the mantle of addressing this issue head-on, demanding rigorous data preprocessing, algorithmic transparency, and the pursuit of fairness.

Through principled design and ethical considerations, responsible AI strives to create systems that reflect the diverse fabric of human society. It urges a concerted effort to bridge digital divides, ensuring that AI's impact is not marred by discriminatory practices. By championing fairness and equity, responsible AI paves the way for a future where technology is a tool of empowerment, rather than an agent of division.

Privacy in the Age of Surveillance: Balancing Innovation and Security

The era of digital advancement has resulted in an unparalleled rise in data creation, raising worries about individual privacy and the security of data. The insatiable appetite of AI for data necessitates a careful equilibrium between creativity and the protection of individual rights in its learning algorithms. Responsible AI highlights the significance of safeguarding data by promoting strong encryption, secure storage, and rigorous access management.

By championing responsible data-handling practices, responsible AI cultivates a sense of trust between technology and individuals. It empowers individuals to retain agency over their personal information while enabling organizations to harness data insights for positive transformations. Thus, it fortifies the pillars of privacy, ensuring that technological advancement does not come at the cost of individual autonomy.

Human-Centric Design: Fostering Collaboration Between Man and Machine

Amidst the AI revolution, the concern that machines will supplant human roles resonates strongly. Responsible AI dispels this notion by embracing a human-centric approach to technology. It envisions AI as an enabler, amplifying human capabilities, enhancing decision making, and fostering innovative synergies between man and machine.

The importance of maintaining human oversight in AI systems cannot be overstated. Responsible AI encourages the development of "explainable AI," wherein the decision-making processes of algorithms are comprehensible and traceable. This not only engenders trust but also empowers individuals to make informed choices, thereby ensuring that AI operates as a benevolent ally rather than an enigmatic force.

Ethics in AI Governance: Navigating a Complex Landscape

Responsible AI extends its purview beyond technology development and encapsulates the intricate realm of governance and regulation. In an era where AI systems traverse legal, social, and cultural boundaries, ensuring ethical oversight becomes paramount. Responsible AI calls for the establishment of robust frameworks, codes of conduct, and regulatory mechanisms that govern the deployment of AI technologies.

The importance of responsible AI governance lies in its ability to avert potential harms, address accountability, and align AI's trajectory with societal aspirations. It prevents a chaotic proliferation of unchecked technology and ensures that AI is wielded for the collective good, ushering in an era of collaborative progress.

Conclusion: The Ongoing Dialogue of Responsibility

As AI embarks on its transformative journey, the importance of responsible AI remains steadfast. It reverberates through technological corridors and resonates in ethical debates, reminding us of the profound influence technology exerts on our lives. The responsibility of shaping AI's trajectory lies in our hands—developers, policymakers, citizens alike—and requires a collective commitment to the tenets of ethical innovation, societal benefit, and accountable stewardship.

In the sections that follow, we navigate deeper into the multidimensional landscape of responsible AI. We unravel its core principles, illuminate real-world applications, and scrutinize its implications on diverse sectors. As we embark on this exploration, we hold the torch of responsibility high, illuminating a path that aligns AI's capabilities with humanity's shared vision for a just, equitable, and ethically enriched future.

Core Ethical Principles

Responsible AI encapsulates a set of guiding principles that govern the ethical development, deployment, and impact of AI technologies. These principles (see Figure 1-1) serve as a compass by which to navigate the intricate intersection of innovation and societal well being. In this summary, we distill the essence of these core principles.

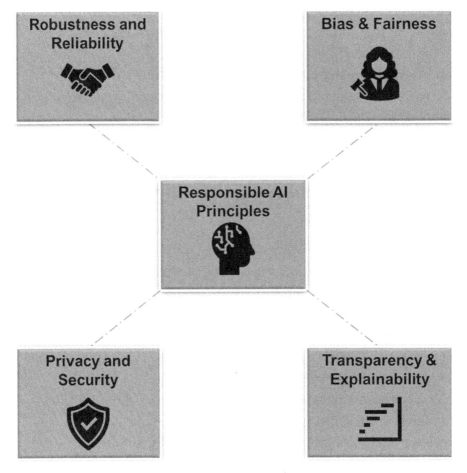

Figure 1-1. *Evolution of artificial intelligence*

1. Bias and Fairness: Cornerstones of Responsible AI

In the realm of AI, the evolution from creativity to ethical obligation has given rise to the notion of responsible AI. Among its guiding principles, Bias and Fairness holds a greater urgency to be tackled than do the others. With the growing integration of AI technologies into our daily lives, assuring the absence of bias and the adherence to fairness principles has risen as a crucial focal point. In this summary, we delve into the intricacies of bias and fairness as foundational elements of responsible AI, exploring their implications and challenges, and the imperative of addressing them in the AI landscape.

Unveiling Bias: The Hidden Challenge

Bias, a deeply ingrained human tendency, can inadvertently seep into AI systems through the data used to train them. AI algorithms learn patterns and associations from vast datasets, which may inadvertently contain biases present in human decisions and societal structures. This can result in discriminatory outcomes, perpetuating stereotypes and exacerbating social disparities.

Responsible AI acknowledges that eliminating all biases may be unfeasible, but mitigating their impact is crucial. The focus shifts to addressing glaring biases that lead to unjust or harmful consequences, while also striving to ensure that AI systems promote equitable treatment for all individuals.

Fairness as a North Star: Ethical Imperative

Fairness in AI underscores the creation of systems that treat all individuals equitably, regardless of their background, demographics, or characteristics. It transcends statistical definitions, delving into ethical considerations to guarantee just and unbiased outcomes. Responsible AI champions fairness as a moral and societal imperative, emphasizing the need to redress historical and systemic inequities.

A critical facet of fairness is algorithmic fairness, which strives to ensure that AI systems' decisions are not influenced by sensitive attributes such as race, gender, or socioeconomic status. Various fairness metrics, algorithms, and techniques have emerged to assess and rectify bias, promoting equitable outcomes and bolstering societal trust in AI technologies.

The Challenge of Quantifying Fairness

The pursuit of fairness encounters challenges in quantification and implementation. Defining a universally acceptable notion of fairness remains elusive, as different contexts demand distinct definitions. Striking a balance between competing notions of fairness poses a significant challenge, with some approaches favoring equal treatment while others prioritize addressing historical disparities.

Quantifying fairness introduces complexities, requiring the calibration of algorithms to meet predefined fairness thresholds. The trade-offs between different types of fairness can be intricate, necessitating careful consideration of their implications for marginalized groups and overall societal well being.

Mitigation Strategies and the Path Forward

Responsible AI advocates for proactive strategies to mitigate bias and ensure fairness in AI systems, as follows:

- Awareness and education play a pivotal role, fostering a deep understanding of biases' manifestation and their potential consequences.

- Data preprocessing techniques, such as re-sampling, re-weighting, and augmentation, offer avenues to alleviate bias in training data.

- Moreover, algorithmic interventions like adversarial training and fairness-aware learning guide AI systems to produce fairer outcomes.

The incorporation of diversity in data collection, model development, and evaluation reduces the risk of perpetuating biases. We will dig deeper into different mitigation strategies in the coming chapters.

Ethical Considerations and Societal Impact

Addressing bias and fostering fairness in AI transcends technical algorithms; it delves into ethical considerations and societal impact. Responsible AI obligates developers, stakeholders, and policymakers to engage in an ongoing dialogue about the ethical dimensions of bias and fairness. It prompts organizations to adopt comprehensive AI ethics frameworks, infusing ethical considerations into the AI development lifecycle.

Societal implications underscore the urgency of addressing bias and promoting fairness. Biased AI systems not only perpetuate existing inequalities but can also erode trust in technology and exacerbate social divisions. By championing fairness, responsible AI cultivates a technological landscape that mirrors society's aspiration for a just and equitable future.

Conclusion: Toward Equitable Technological Frontiers

In the pursuit of responsible AI, addressing bias and ensuring fairness is not a mere checkbox; it is a transformative endeavor that demands collaboration, ingenuity, and ethical conviction. As AI technologies continue to reshape industries and touch countless lives, upholding the principles of bias mitigation and fairness is an ethical imperative. The path forward involves a multidisciplinary approach, where technological innovation converges with ethical considerations, paving the way for a future where AI fosters inclusivity, equity, and the betterment of humanity as a whole.

2. Transparency and Explainability

In the realm of AI, where algorithms make decisions that impact our lives, the principles of transparency and explainability emerge as critical safeguards. These principles are integral components of responsible AI, a framework designed to ensure ethical, fair, and accountable AI development and deployment. In this summary, we explore the significance of transparency and explainability as cornerstones of responsible AI, delving into their implications, challenges, and the transformative potential they offer.

Transparency: Illuminating the Black Box

Transparency in AI refers to the openness and comprehensibility of an AI system's decision-making process. It addresses the "black box" nature of complex AI algorithms, where inputs and processes result in outputs, without clear visibility into the reasoning behind those outcomes. Responsible AI demands that developers and stakeholders make AI systems transparent, enabling individuals to understand how decisions are reached.

Transparency serves multiple purposes. It fosters accountability, allowing developers to identify and rectify biases, errors, or unintended consequences. It also empowers individuals affected by AI decisions to challenge outcomes that seem unfair or discriminatory. Moreover, transparency cultivates trust between AI systems and users, a crucial element for widespread adoption.

However, achieving transparency is no trivial task. AI models often consist of intricate layers and nonlinear transformations, making it challenging to extract human-interpretable insights. Balancing the need for transparency with the complexity of AI algorithms remains a delicate endeavor.

Explainability: Bridging the Gap

Explainability complements transparency by providing insights into the rationale behind AI decisions in a human-understandable manner. While transparency reveals the overall decision-making process, explainability delves into the specifics, unraveling the factors that contributed to a particular outcome.

Explainability addresses the cognitive gap between the inherently complex nature of AI processes and human cognition. It strives to answer questions like "Why was this decision made?" and "How did the algorithm arrive at this conclusion?" By translating AI outputs into explanations that resonate with human reasoning, explainability empowers users to trust and engage with AI systems more confidently.

However, achieving meaningful explainability is not without its challenges. Striking a balance between simplicity and accuracy, especially in complex AI models like deep neural networks, requires innovative techniques that synthesize complex interactions into interpretable insights.

Implications and Applications

The implications of transparency and explainability extend across a spectrum of AI applications. In sectors like finance and health care, where AI-driven decisions can have profound consequences, transparency and explainability help stakeholders understand risk assessments, diagnoses, and treatment recommendations. In the criminal justice system, these principles can ensure that AI-driven predictive models do not perpetuate racial or socioeconomic biases.

Furthermore, transparency and explainability are essential for regulatory compliance. As governments and institutions craft AI governance frameworks, having the ability to audit and verify AI decisions becomes pivotal. Transparent and explainable AI systems enable regulators to assess fairness, accuracy, and compliance with legal and ethical standards.

Challenges and Future Directions

While the importance of transparency and explainability is widely recognized, challenges persist, such as the following:

- The trade-off between model complexity and interpretability remains a fundamental conundrum. Developing techniques that maintain accuracy while providing clear explanations is an ongoing research frontier.

- The dynamic nature of AI models also poses challenges. Explainability should extend beyond initial model deployment to cover model updates, adaptations, and fine-tuning. Ensuring explanations remain accurate and meaningful throughout an AI system's lifecycle is a complex task.

- Moreover, balancing transparency with proprietary considerations is a delicate tightrope walk. Companies may be reluctant to reveal proprietary algorithms or sensitive data, but striking a balance between intellectual property protection and the public's right to transparency is imperative.

Conclusion

Transparency and explainability are not mere technical prerequisites but rather essential pillars of responsible AI. They foster trust, accountability, and informed decision making in an AI-driven world. By shedding light on AI's decision-making processes and bridging the gap between algorithms and human understanding, transparency and explainability lay the foundation for an ethical, fair, and inclusive AI landscape. As AI continues to evolve, embracing these principles ensures that the journey into the future is guided by clarity, integrity, and empowerment.

3. Privacy and Security

In the age of rapid technological advancement, the integration of AI into various facets of our lives brings with it a plethora of benefits and challenges. As AI systems become more widely used and more complex, the preservation of individual privacy and the assurance of data security emerge as crucial facets within the scope of responsible AI. This summary delves into the intricate interplay between privacy and security, outlining their significance, implications, and the pivotal role they play as core principles in the responsible development and deployment of AI technologies.

Privacy in the Digital Age: A Precious Commodity

Privacy, a cornerstone of personal freedom, takes on new dimensions in the digital era. As AI systems accumulate vast amounts of data for analysis and decision making, the preservation of individuals' rights to privacy becomes paramount. Responsible AI recognizes the necessity of preserving privacy as an inherent human entitlement, guaranteeing that personal information is managed with the highest level of consideration and reverence.

One of the key tenets of responsible AI is informed consent. Individuals have the right to know how their data will be used and shared, granting them the agency to make informed decisions. Transparent communication between AI developers and users fosters a sense of trust and empowers individuals to maintain control over their personal information.

Furthermore, data minimization is a fundamental principle underpinning responsible AI. It advocates for the collection, processing, and retention of only that data essential for a specific AI task. This approach minimizes the risk of unintended exposure and helps mitigate potential breaches of privacy.

Data Security: Fortifying the Digital Fortress

The inseparable companion of privacy is data security. Responsible AI recognizes that the data collected and utilized by AI systems is a valuable asset, and safeguarding it against unauthorized access, manipulation, or theft is imperative. Robust data security measures, including encryption, access controls, and secure storage, form the backbone of a trustworthy AI ecosystem.

Responsible AI developers must prioritize data protection throughout the AI life cycle. From data collection and storage to data sharing and disposal, security protocols must be rigorously implemented. By fortifying the digital fortress, responsible AI endeavors to shield sensitive information from malicious intent, preserving the integrity of individuals' identities and experiences.

Challenges and Opportunities

While privacy and security stand as cornerstones of responsible AI, they also present intricate challenges that demand innovative solutions, as follows:

- The vast quantities of data collected by AI systems necessitate sophisticated anonymization techniques to strip away personal identifiers, ensuring that individuals' privacy is upheld even in aggregated datasets.

- Additionally, the global nature of dataflows necessitates a harmonized approach to privacy and security regulations. Responsible AI advocates for the establishment of international standards that guide data-handling practices, transcending geographical boundaries and ensuring consistent protection for individuals worldwide.

In the face of these challenges, responsible AI opens doors to transformative opportunities. Privacy-preserving AI techniques, such as federated learning and homomorphic encryption, empower AI systems to learn and generate insights from decentralized data sources without compromising individual privacy. These innovative approaches align with the ethos of responsible AI, fostering both technological progress and ethical integrity.

Trust and Beyond: The Nexus of Privacy, Security, and Responsible AI

The interweaving of privacy and security within the fabric of responsible AI extends far beyond technical considerations. It is an embodiment of the ethical responsibility that AI developers and stakeholders bear toward individuals whose data fuels the AI ecosystem. By prioritizing privacy and security, responsible AI cultivates trust between technology and humanity, reinforcing the societal acceptance and adoption of AI technologies.

Responsible AI acknowledges that the preservation of privacy and security is not a matter of mere regulatory compliance, but rather one of ethical duty. It encompasses the commitment to treat data as a stewardship, handling it with integrity and ensuring that it is leveraged for the collective good rather than misused for unwarranted surveillance or exploitation.

In conclusion, privacy and security emerge as inseparable twins within the constellation of responsible AI principles. Their significance extends beyond the technological realm, embodying the ethical foundation upon which AI technologies stand. By embracing and upholding these principles, responsible AI charts a path toward a future where technological advancement and individual rights coexist harmoniously, empowering society with the transformative potential of AI while safeguarding the sanctity of privacy and data security.

4. Robustness and Reliability

In the ever evolving landscape of AI, ensuring the robustness and reliability of AI systems stands as a paramount principle of responsible AI. Robustness entails the ability of AI models to maintain performance across diverse and challenging scenarios, while reliability demands the consistent delivery of accurate outcomes. This summary delves into the significance of these intertwined principles, their implications, and the measures essential for their realization within AI systems.

Robustness: Weathering the Storms of Complexity

Robustness in AI embodies the capacity of algorithms to remain effective and accurate amidst complexity, uncertainty, and adversarial conditions. AI systems that lack robustness may falter when confronted with novel situations, data variations, or deliberate attempts to deceive them. A robust AI model can adeptly generalize from its training data to novel, real-world scenarios, minimizing the risk of errors and biases that could undermine its utility and trustworthiness.

The importance of robustness resonates across numerous domains. In self-driving cars, a robust AI system should reliably navigate various weather conditions, road layouts, and unexpected obstacles. In medical diagnostics, robust AI models ensure consistent accuracy across diverse patient profiles and medical settings. Addressing the challenges of robustness is crucial to building AI systems that excel in real-world complexity.

Reliability: A Pillar of Trust

Reliability complements robustness by emphasizing the consistent delivery of accurate outcomes over time. A reliable AI system maintains its performance not only under challenging conditions but also through continuous operation. Users, stakeholders, and society as a whole rely on AI systems for critical decisions, making reliability a foundational element of trust.

Unreliable AI systems can lead to dire consequences. In sectors such as finance, where AI aids in risk assessment and investment strategies, an unreliable model could lead to substantial financial losses. In health care, an unreliable diagnostic AI could compromise patient well being. The pursuit of reliability ensures that AI consistently upholds its performance standards, enabling users to confidently integrate AI-driven insights into their decision-making processes.

Challenges and Mitigation Strategies

Achieving robustness and reliability in AI systems is no small feat, as these principles intersect with multiple dimensions of AI development and deployment, as follows:

- Data Quality and Diversity: Robust AI requires diverse and representative training data that encompass a wide array of scenarios. Biased or incomplete data can undermine robustness. Responsible AI emphasizes data quality assurance, unbiased sampling, and continuous monitoring to ensure that AI models learn from a comprehensive dataset.

- Adversarial Attacks: AI systems vulnerable to adversarial attacks can make erroneous decisions when exposed to subtly altered input data. Defending against such attacks involves robust training strategies, adversarial training, and constant model evaluation to fortify AI systems against potential vulnerabilities.

- Transfer Learning and Generalization: The ability to generalize knowledge from one domain to another is crucial for robustness. AI developers employ transfer learning techniques to ensure that models can adapt and perform well in new contexts without extensive retraining.

- Model Monitoring and Feedback Loops: To ensure reliability, continuous monitoring of AI models in real-world scenarios is imperative. Feedback loops allow models to adapt and improve based on their performance, enhancing reliability over time.

- Interpretable AI: Building AI systems that provide transparent insights into their decision-making processes enhances both robustness and reliability. Interpretable AI empowers users to understand and trust AI-generated outcomes, fostering reliability in complex decision domains.

- Collaborative Ecosystems: The collaborative efforts of researchers, developers, policymakers, and domain experts are vital for advancing robust and reliable AI. Open dialogues, knowledge sharing, and interdisciplinary cooperation facilitate the identification of challenges and the development of effective mitigation strategies.

Conclusion: Building Bridges to Trustworthy AI

Robustness and reliability stand as pillars of responsible AI, nurturing the growth of AI systems that can navigate complexity, deliver accurate results, and engender trust. In the pursuit of these principles, AI practitioners tread a path of continuous improvement, where technological advancement intertwines with ethical considerations. As AI takes on an increasingly pivotal role in our lives, the pursuit of robustness and reliability ensures that it remains a tool of empowerment, enhancing human endeavors across sectors while safeguarding the foundations of trust and accountability.

Conclusion

In this chapter, we have provided a comprehensive introduction to the world of AI, highlighting its immense potential to transform industries and societies. We delved into the imperative need for responsible AI, acknowledging that as AI's influence grows, so too does the significance of ensuring its ethical and moral dimensions. By examining the core principles of responsible AI, such as fairness, transparency, security, and reliability, we've underscored the essential framework for guiding AI development and deployment.

It is clear that responsible AI is not merely an option, but also an ethical obligation that ensures technology serves humanity in a just and equitable manner. As we move forward, embracing responsible AI will be pivotal in shaping a future where innovation and ethical considerations harmoniously coexist. In the next chapters, we will deep-dive into each of the core principles and showcase how they can be achieved through sample examples with code walkthroughs.

CHAPTER 2

Bias and Fairness

In the artificial intelligence (AI) landscape, bias's impact on decisions is paramount. From individual choices to complex models, bias distorts outcomes and fairness. Grasping bias's nuances is essential for equitable systems. It's a complex interplay of data and beliefs with profound implications. Detecting and mitigating bias is empowered by technology, nurturing transparent and responsible AI. This ongoing quest aligns with ethics, sculpting AI that champions diversity and societal progress.

In this chapter, we delve into the intricate relationship between bias, fairness, and artificial intelligence. We explore how bias can impact decision making across various domains, from individual judgments to automated systems. Understanding the types and sources of bias helps us identify its presence in data and models. We also delve into the importance of recognizing bias for creating fair and equitable systems and how explainable AI aids in this process. Additionally, we touch on techniques to detect, assess, and mitigate bias, as well as the trade-offs between model complexity and interpretability. This comprehensive exploration equips us to navigate the complexities of bias and fairness in the AI landscape, fostering ethical and inclusive AI systems.

Understanding Bias in Data and Models

Bias in data and models refers to the presence of systematic deviations that lead to inaccuracies or unfairness in decision-making processes. It emerges when data-collection or model-construction processes

© Avinash Manure, Shaleen Bengani, Saravanan S 2023
A. Manure et al., *Introduction to Responsible AI*,
https://doi.org/10.1007/978-1-4842-9982-1_2

inadvertently favor certain groups, attributes, or perspectives over others. This bias can stem from various sources, such as historical inequalities, flawed data-collection methods, or biased algorithms. Addressing bias requires a deep understanding of its manifestations in both data and model outcomes, along with the implementation of strategies that ensure equitable and unbiased decision making in artificial intelligence systems.

Importance of Understanding Bias

Understanding bias is an indispensable cornerstone when striving to establish systems that are fair and equitable, especially within the domain of artificial intelligence (AI) and machine learning (ML). Bias holds the potential to instigate systemic inequalities, perpetuate discrimination, and reinforce social disparities. Recognizing and comprehending its significance is paramount for fostering inclusivity, upholding ethical practices, and ensuring that AI technologies make a positive contribution to society. Delving deeper, let's explore the profound importance of understanding bias for the creation of fair and equitable systems, beginning with the following:

- **Avoiding Discrimination and Inequity:** Bias, whether embedded within data or woven into models, can be the catalyst for generating discriminatory outcomes. In instances where AI systems are crafted without careful consideration of bias, they run the risk of disproportionately disadvantaging specific groups, thereby perpetuating pre-existing inequalities. A profound understanding of the origins of bias and its far-reaching implications empowers developers to embark on the journey of crafting systems that treat all individuals impartially, irrespective of factors like background, gender, race, or any other defining characteristic.

- **Ensuring Ethical AI Deployment:** Ethics and responsibility form the bedrock of AI development endeavours. The comprehension of bias equips developers with the capability to align their work with ethical principles and legal mandates. The essence of ethical AI lies in its determination to steer clear of accentuating or prolonging societal biases. Instead, ethical AI strives ardently for fairness, transparency, and accountability, serving as a beacon of responsible technological advancement.

- **Building Trust in AI Systems:** The acceptability and trustworthiness of AI hinge upon its perceived fairness and impartiality. An AI system that consistently generates biased outcomes erodes public trust and undermines confidence in its efficacy. By proactively addressing bias in its myriad forms, developers embark on a journey to construct systems that radiate trustworthiness and credibility, reinforcing the belief that AI technologies are designed to function equitably.

- **Enhancing Decision-Making Processes:** AI systems are progressively integrated into decision-making processes that wield a tangible impact on individuals' lives—be it hiring, lending, or criminal justice. Bias within these systems can give rise to outcomes that are unjust and inequitable. Herein lies the critical role of understanding bias: it lays the foundation for AI-driven decisions that are well informed, transparent, and free from any semblance of discriminatory influence.

- **Promoting Innovation:** Bias possesses the potential to shackle AI systems, limiting their efficacy and applicability. A system tainted by bias may fail to accurately represent the diverse spectrum of human experiences and perspectives. Addressing bias serves as a catalyst for innovation, creating an environment conducive to the development of AI systems that are adaptive, versatile, and potent across various contexts.

- **Reducing Reproduction of Historical Injustices:** The shadows of historical biases and injustices can unwittingly find their way into AI systems that learn from biased data. In this context, understanding these latent biases proves instrumental. It empowers developers to take proactive measures, preventing AI from inadvertently perpetuating negative historical patterns and detrimental stereotypes.

- **Encouraging Diversity and Inclusion:** Understanding bias emerges as a driving force behind fostering diversity and inclusion within the realm of AI development. By acknowledging biases and their potential impact, developers take on the responsibility of ensuring that their teams are a microcosm of diversity, ushering in an array of perspectives that contribute to more-comprehensive system design and judicious decision making.

- **Contributing to Social Progress:** AI possesses an immense potential to be a conduit of positive transformation, capable of precipitating societal progress. Through the lens of addressing bias and architecting fair systems, AI emerges as a tool that can bridge disparities, champion equal opportunities, and propel social aspirations forward.

- **Long-Term Viability of AI:** With AI poised to permeate diverse sectors, ranging from health care to education to finance, the need for long-term viability and sustainable adoption becomes evident. This enduring viability is anchored in the creation of AI technologies that are inherently equitable, acting as catalysts for positive change and responsible technological advancement.

Understanding bias extends beyond theoretical recognition; it serves as a guiding beacon that informs ethical practices, shapes technological landscapes, and steers the trajectory of AI's contribution to society.

How Bias Can Impact Decision-Making Processes

Bias can have a profound impact on decision-making processes across various domains, from individual judgments to complex automated systems. It can distort perceptions, influence choices, and lead to unjust or discriminatory outcomes. Understanding how bias affects decision making is crucial for developing fair and equitable systems. Here's an in-depth explanation of how bias can impact decision-making processes:

- **Distorted Perceptions:** Bias can alter how information is perceived and interpreted. When bias is present, individuals may focus more on certain aspects of a situation while overlooking others. This can lead to incomplete or skewed understandings, ultimately influencing the decisions made.

- **Unconscious Biases:** Human decision making is influenced by unconscious biases, often referred to as implicit biases. These biases stem from cultural, societal, and personal factors and can unconsciously shape perceptions, attitudes, and judgments. Even well-intentioned individuals can be impacted by these biases without realizing it.

- **Confirmation Bias:** Confirmation bias occurs when individuals seek out or favor information that confirms their existing beliefs or biases. This can result in decisions that are not well informed or balanced, as contradictory information may be ignored or dismissed.

- **Stereotyping:** Bias can lead to stereotyping, where individuals make assumptions about a person or group based on preconceived notions. Stereotyping can result in decisions that are unfair, as they are based on generalizations rather than individual merits.

- **Unequal Treatment:** Bias can lead to unequal treatment of different individuals or groups. This can manifest in various ways, such as offering different opportunities, resources, or punishments based on factors like race, gender, or socioeconomic status.

- **Discriminatory Outcomes:** When bias influences decisions, it can lead to discriminatory outcomes. Discrimination can occur at both individual and systemic levels, affecting people's access to education, employment, health care, and more.

- **Impact on Automated Systems:** In automated decision-making systems, bias present in training data can lead to biased predictions and recommendations. These systems may perpetuate existing biases and further entrench inequality if not properly addressed.

- **Feedback Loops:** Biased decisions can create feedback loops that perpetuate and amplify bias over time. For example, if biased decisions lead to limited opportunities for a particular group, it can reinforce negative stereotypes and further marginalize that group.

- **Erosion of Trust:** When individuals perceive that decision-making processes are influenced by bias, it erodes trust in those processes and the institutions responsible for them. This can lead to social unrest and a breakdown of societal cohesion.

- **Reinforcing Inequalities:** Bias in decision making can reinforce existing social inequalities. If certain groups consistently face biased decisions, their opportunities and access to resources are limited, perpetuating a cycle of disadvantage.

Types of Bias

Bias in machine learning refers to the presence of systematic and unfair errors in data or models that can lead to inaccurate or unjust predictions, decisions, or outcomes. There are several types of bias that can manifest in different stages of the machine learning pipeline (see Figure 2-1).

Data	Model	Social

- Sampling Bias
- Measurement Bias
- Coverage Bias

- Representation Bias
- Algorithmic Bias
- Feedback Loop Bias

- Cultural Bias
- Gender Bias
- Economic Bias

Figure 2-1. *Types of bias*

1. **Data Bias:** Data bias encompasses biases present in the data used to train and test machine learning models. This bias can arise due to various reasons, such as the following:

   - **Sampling Bias:** When the collected data is not representative of the entire population, leading to over- or under-representation of certain groups or attributes. For instance, in a medical diagnosis dataset, if only one demographic group is represented, the model might perform poorly for underrepresented groups.

   - **Measurement Bias:** Errors or inconsistencies introduced during data-collection or measurement processes can introduce bias. For example, if a survey is conducted in a language not understood by a specific community, their perspectives will be omitted, leading to biased conclusions.

   - **Coverage Bias:** Occurs when certain groups or perspectives are missing from the dataset. This can result from biased data-collection methods, incomplete sampling, or systemic exclusion.

2. **Model Bias:** Model bias emerges from the learning algorithms' reliance on biased data during training, which can perpetuate and sometimes amplify biases, as follows:

- **Representation Bias:** This occurs when the features or attributes used for training disproportionately favor certain groups. Models tend to learn from the biases present in the training data, potentially leading to biased predictions.

- **Algorithmic Bias:** Some machine learning algorithms inherently perpetuate biases. For example, if a decision-tree algorithm learns to split data based on biased features, it will reflect those biases in its predictions.

- **Feedback Loop Bias:** When models' predictions influence real-world decisions that subsequently affect the data used for future training, a feedback loop is created. Biased predictions can perpetuate over time, reinforcing existing biases.

3. **Social Bias:** Social bias pertains to the biases present in society that get reflected in data and models, as follows:

- **Cultural Bias:** Cultural norms, beliefs, and values can shape how data is collected and interpreted, leading to biased outcomes.

- **Gender Bias:** Historical and societal gender roles can result in unequal representation in datasets, affecting model performance.

- **Racial Bias:** Biased historical practices can lead to underrepresentation or misrepresentation of racial groups in data, impacting model accuracy.

- **Economic Bias:** Socioeconomic disparities can lead to differences in data availability and quality, influencing model outcomes.

Understanding these types of bias is essential for developing strategies to detect, mitigate, and prevent bias. Addressing bias involves a combination of careful data collection, preprocessing, algorithm selection, and post-processing interventions. Techniques such as reweighting, resampling, and using fairness-aware algorithms can help mitigate bias at various stages of model development.

However, ethical considerations play a crucial role in addressing bias. Being aware of the potential impact of bias on decision-making processes and actively working to mitigate it aligns AI development with principles of fairness, transparency, and accountability. By understanding the different types of bias, stakeholders can work toward creating AI systems that promote equitable outcomes across diverse contexts and populations.

Examples of Real-world Cases Where Models Exhibited Biased Behavior

Several real-world examples illustrate how machine learning models have exhibited biased behavior, leading to unfair and discriminatory outcomes. These cases highlight the importance of addressing bias in AI systems to avoid perpetuating inequality and to ensure ethical and equitable deployments. Here are some detailed examples:

1. **Amazon's Gender-Biased Hiring Tool:** In 2018, it was revealed that Amazon had developed an AI-driven recruiting tool to help identify top job

candidates. However, the system displayed a bias against female applicants. This bias resulted from the training data, which predominantly consisted of resumes submitted over a ten-year period, mostly from male candidates. As a result, the model learned to favor male applicants and downgrade resumes that included terms associated with women.

2. **Racial Bias in Criminal Risk Assessment:** Several criminal risk assessment tools used in the criminal justice system have been criticized for exhibiting racial bias. These tools predict the likelihood of reoffending based on historical arrest and conviction data. However, the historical bias in the data can lead to overestimating the risk for minority groups, leading to discriminatory sentencing and parole decisions.

3. **Google Photos' Racist Labeling:** In 2015, Google Photos' auto-tagging feature was found to label images of Black people as "gorillas." This was a result of the model's biased training data, which did not include enough diverse examples of Black individuals. The incident highlighted the potential harm of biased training data and the need for inclusive datasets.

4. **Biased Loan Approval Models:** Machine learning models used for loan approval have shown bias in favor of certain demographic groups. Some models have unfairly denied loans to minority applicants or offered them higher interest rates, reflecting historical biases in lending data.

5. **Facial Recognition and Racial Bias:** Facial
    recognition systems have been criticized for their
    racial bias, where they are more likely to misidentify
    people with darker skin tones, particularly women.
    This bias can result in inaccurate surveillance, racial
    profiling, and infringement of civil rights.

These real-world examples underscore the urgency of addressing bias
in AI systems. To prevent such biased behavior, it's crucial to carefully
curate diverse and representative training data, use fairness-
aware algorithms, implement bias detection and mitigation techniques,
and continuously monitor and evaluate model outputs for fairness.
By proactively addressing bias, developers can ensure that AI systems
contribute positively to society and uphold ethical standards.

Techniques to Detect and Mitigate Bias

Detecting and mitigating bias in machine learning models and data
is essential to create fair and equitable AI systems. Let's look at some
techniques to identify and address bias (Figure 2-2).

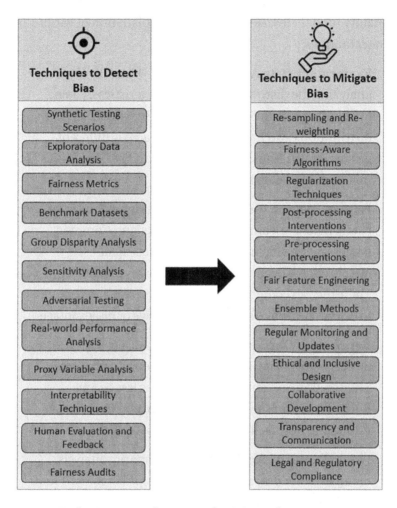

Figure 2-2. *Techniques to detect and mitigate bias*

Techniques to Detect Bias

Bias-detection techniques are essential tools for identifying and quantifying biases present in data, models, and their outputs. These techniques help ensure that AI systems are fair, equitable, and free from discriminatory tendencies. Here's a detailed explanation of various bias-detection techniques:

- **Exploratory Data Analysis (EDA):** EDA involves analyzing the distribution and characteristics of data to identify potential sources of bias. By visualizing data distributions and exploring patterns across different groups or attributes, data scientists can spot disparities that might indicate bias.

- **Fairness Metrics:** Fairness metrics quantify and measure bias in machine learning models' predictions. Common fairness metrics include disparate impact, equal opportunity difference, and statistical parity difference. These metrics compare outcomes between different groups to determine if there's an unfair advantage or disadvantage.

- **Benchmark Datasets:** Benchmark datasets are designed to expose bias in machine learning models. They contain examples where fairness issues are intentionally present, making them useful for evaluating how well models handle bias.

- **Group Disparity Analysis:** Group disparity analysis compares outcomes for different groups across various attributes. By calculating differences in outcomes, such as acceptance rates, loan approvals, or hiring decisions, developers can identify disparities that indicate bias.

- **Sensitivity Analysis:** Sensitivity analysis involves testing how small changes in data or model inputs impact outcomes. This can reveal how sensitive predictions are to variations in the input, helping identify which features contribute most to biased outcomes.

- **Adversarial Testing:** Adversarial testing involves deliberately introducing biased data or biased inputs to observe how models respond. By observing how models react to these adversarial inputs, developers can gauge their susceptibility to bias.

- **Real-world Performance Analysis:** Deployed models can be monitored in real-world settings to assess whether they generate biased outcomes in practice. Continuous monitoring allows developers to detect emerging bias patterns over time.

- **Proxy Variable Analysis:** Proxy variables are attributes correlated with protected characteristics (e.g., gender, race). Analyzing how strongly proxy variables affect model outcomes can indicate the presence of hidden bias.

- **Interpretability Techniques:** Interpretability techniques, like feature importance analysis, can help understand which features contribute most to model predictions. Biased features that contribute disproportionately might indicate bias.

- **Human Evaluation and Feedback:** Involving human evaluators from diverse backgrounds to review model outputs and provide feedback can help identify bias that might not be apparent through automated techniques.

- **Fairness Audits:** Fairness audits involve a comprehensive review of data collection, preprocessing, and model development processes to identify potential sources of bias.

- **Synthetic Testing Scenarios:** Creating controlled scenarios with synthetic data can help simulate potential bias sources to observe their impact on model predictions.

Techniques to Mitigate Bias

Mitigating bias in machine learning models is a critical step to ensure fairness and equitable outcomes. There are various strategies and techniques that can be employed to reduce bias and promote fairness in AI systems. Here's a detailed explanation of mitigation bias strategies:

1. **Resampling:** Balancing class representation by either oversampling underrepresented groups or undersampling overrepresented ones can help reduce bias present in the data.

2. **Reweighting:** Assigning different weights to different classes or samples can adjust the model's learning process to address imbalances.

3. **Fairness-Aware Algorithms:**

   - **Adversarial Debiasing:** Incorporates an additional adversarial network to reduce bias while training the main model, forcing it to disregard features correlated with bias.

   - **Equalized Odds:** Adjusts model thresholds to ensure equal opportunity for positive outcomes across different groups.

- **Reject Option Classification:** Allows the model to decline to make a prediction when uncertainty about its fairness exists.

4. **Regularization Techniques:**

   - **Fairness Constraints:** Adding fairness constraints to the model's optimization process to ensure predictions are within acceptable fairness bounds.

   - **Lagrangian Relaxation:** Balancing fairness and accuracy trade-offs by introducing Lagrange multipliers during optimization.

5. **Post-processing Interventions:**

   - **Calibration:** Adjusting model predictions to align with desired fairness criteria while maintaining overall accuracy.

   - **Reranking:** Reordering model predictions to promote fairness without significantly compromising accuracy.

6. **Preprocessing Interventions:**

   - **Data Augmentation:** Adding synthesized data points to underrepresented groups to improve model performance and reduce bias.

   - **De-biasing Data Preprocessing:** Using techniques like reweighting or resampling during data preprocessing to mitigate bias before training.

7. **Fair Feature Engineering:** Creating or selecting features that are less correlated with bias, which can help the model focus on relevant and fair attributes.

8. **Ensemble Methods:** Combining multiple models that are trained with different strategies can help mitigate bias, as biases in individual models are less likely to coincide.

9. **Regular Monitoring and Updates:** Continuously monitoring model performance for bias in real-world scenarios and updating the model as new data becomes available to ensure ongoing fairness.

10. **Ethical and Inclusive Design:** Prioritizing diverse representation and ethical considerations in data collection, preprocessing, and model development to prevent bias from entering the system.

11. **Collaborative Development:** Involving stakeholders from diverse backgrounds, including ethicists and affected communities, to collaboratively address bias and ensure that mitigation strategies align with ethical values.

12. **Transparency and Communication:** Being transparent about the steps taken to mitigate bias and communicating these efforts to users and stakeholders to build trust in the system.

13. **Legal and Regulatory Compliance:** Ensuring that the AI system adheres to relevant laws and regulations concerning discrimination and bias, and actively working to comply with them.

Implementing Bias Detection and Fairness

The purpose of this exercise was to start exploring bias, including potential methods to reduce bias as well as how bias may very easily become exacerbated in ML models. In this exercise, for the sake of brevity, Let's check for bias in relation to race, though bias should be checked against the other protected classes as well.

Stage 1: Data Bias

In this task, we begin by training a model on the original dataset. However, upon analyzing and visualizing the data, we detect the presence of racial bias. To address this issue, we implement a resampling technique to promote a fair and unbiased representation.

Resampling is a technique used in machine learning to create new training data by altering the original data. It can include both oversampling and undersampling and aims to create a more balanced and representative training dataset, which helps models learn more effectively.

We then retrain the model on the balanced data and evaluate the accuracy. This process helps to mitigate race bias and ensure fair prediction.

Dataset Details

We have used an individual's annual income results from various factors. Intuitively, it is influenced by the individual's education level, age, gender, occupation, etc.

Source: https://archive.ics.uci.edu/dataset/2/adult

The dataset contains the following 16 columns:

- **Age:** Continuous

- **Workclass:** Private, Self-emp-not-inc, Self-emp-inc, Federal-gov, Local-gov, State-gov, Without-pay, Never-worked

- **Fnlwgt:** Continuous

- **Education:** Bachelor's, Some-college, 11th, HS-grad, Prof-school, Assoc-acdm, Assoc-voc, 9th, 7th-8th, 12th, Master's, 1st-4th, 10th, Doctorate, 5th-6th, Preschool

- **Marital Status:** Married-civ-spouse, Divorced, Never-married, Separated, Widowed, Married-spouse-absent, Married-AF-spouse

- **Occupation:** Tech-support, Craft-repair, Other-service, Sales, Exec-managerial, Prof-specialty, Handlers-cleaners, Machine-op-inspect, Adm-clerical, Farming-fishing, Transport-moving, Priv-house-serv, Protective-serv, Armed-Forces

- **Relationship:** Wife, Own-child, Husband, Not-in-family, Other-relative, Unmarried

- **Race:** White, Asian-Pac-Islander, Amer-Indian-Eskimo, Other, Black

- **Sex:** Female, Male

- **Capital Gain:** Continuous

- **Capital Loss:** Continuous

- **Hours-per-week:** Continuous

- **Native Country:** United-States, Cambodia, England, Puerto-Rico, Canada, Germany, Outlying-US, India, Japan, Greece, South, China, Cuba, Iran, Honduras, Philippines, Italy, Poland, Jamaica, Vietnam, Mexico, Portugal, Ireland, France, Dominican-Republic, Laos, Ecuador, Taiwan, Haiti, Columbia, Hungary, Guatemala, Nicaragua, Scotland, Thailand, Yugoslavia, El-Salvador, Trinadad&Tobago, Peru, Hong, Holand-Netherlands

- **Income** (>50k or <=50k): Target variable

Getting Started

The following is the process for implementing data bias detection and mitigation process in Python.

Step 1: Importing Packages

The following shows how to import all the necessary packages:

```
[In]:
# Import necessary libraries
import pandas as pd
import numpy as np
from sklearn.model_selection import train_test_split
from sklearn.ensemble import RandomForestClassifier
from sklearn.metrics import accuracy_score, classification_
report, confusion_matrix
from sklearn.utils import resample
from sklearn.preprocessing import LabelEncoder, StandardScaler
from sklearn.metrics import classification_report
```

Step 2: Loading the Data

```
[In]:
# Read the dataset into a pandas DataFrame
df = pd.read_csv(" Income.csv")
```

Step 3: Checking the Data Characteristics

Check if there are any discrepancies in the data, like missing values, wrong data types, etc.:

```
[In]:
# Display basic information about the dataset
df.info()
```

```
[Out]:
<class 'pandas.core.frame.DataFrame'>
RangeIndex: 48842 entries, 0 to 48841
Data columns (total 15 columns):
 #   Column          Non-Null Count  Dtype
---  ------          --------------  -----
 0   age             48842 non-null  int64
 1   workclass       48842 non-null  int32
 2   fnlwgt          48842 non-null  int64
 3   education       48842 non-null  int32
 4   education-num   48842 non-null  int64
 5   marital-status  48842 non-null  int32
 6   occupation      48842 non-null  int32
 7   relationship    48842 non-null  int32
 8   race            48842 non-null  int32
 9   sex             48842 non-null  int32
 10  capital-gain    48842 non-null  int64
 11  capital-loss    48842 non-null  int64
 12  hours-per-week  48842 non-null  int64
```

```
13   native-country   48842 non-null   int32
14   income            48842 non-null   int32
dtypes: int32(9), int64(6)
memory usage: 3.9 MB
```

There are no null values present in the data, so we can proceed with the data preprocessing steps.

Step 4: Data Preprocessing

Create a list of columns with categorical columns to be encoded:

```
[In]:
# Define a list of categorical columns to be encoded and
perform label encoding for categorical columns
categorical_columns = ['sex', 'race', 'education', 'marital-
status', 'occupation', 'relationship', 'native-country',
'workclass', 'income']
label_encoders = {}
for column in categorical_columns:
    label_encoders[column] = LabelEncoder()
    df[column] = label_encoders[column].fit_
transform(df[column])
```

Categorical columns contain multiple categorical values. To use these categorical values for model building, apply dummy variable-creation techniques to columns having more than two unique values.

```
[In]:
# Perform one-hot encoding for columns with more than 2
categories
get_dummies = []
label_encoding = []
```

```
for i in categorical_columns:
    print('Column Name:', i, ', Unique Value Counts:',
len(df[i].unique()), ', Values:', df[i].unique())
    if len(df[i].unique()) > 2:
        get_dummies.append(i)
    else:
        label_encoding.append(i)
df = pd.get_dummies(df, prefix=get_dummies, columns=get_dummies)
```

```
[Out]:
Column Name: sex, Unique Value Counts: 2, Values: [1 0]
Column Name: race, Unique Value Counts: 2, Values: [1 0]
Column Name: education, Unique Value Counts: 16, Values: [ 9
11  1 12  6 15  7  8  5 10 14  4  0  3 13  2]
Column Name: marital-status, Unique Value Counts: 7, Values: [4
2 0 3 5 1 6]
Column Name: occupation, Unique Value Counts: 15, Values: [
1  4  6 10  8 12  3 14  5  7 13  0 11  2  9]
Column Name: relationship, Unique Value Counts: 6, Values: [1 0
5 3 4 2]
Column Name: native-country, Unique Value Counts: 42, Values:
[39  5 23 19  0 26 35 33 16  9  2 11 20 30 22 31  4  1 37  7 25
36 14 32
  6  8 10 13  3 24 41 29 28 34 38 12 27 40 17 21 18 15]
Column Name: workclass, Unique Value Counts: 9, Values: [7 6 4
1 2 0 5 8 3]
Column Name: income, Unique Value Counts: 2, Values: [0 1]
```

```
[In]:
# Gender distribution graph
df['sex'].value_counts().plot(kind='bar')
```

```
[Out]:
```

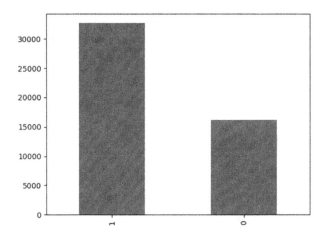

Figure 2-3. *Gender distribution, male vs. female*

As shown in Figure 2-3, with 67% of the population identified as male and 33% as female, which is considered as imbalanced dataset in the context of machine learning. After comparing both gender and demographic features in the dataset, its more critical to prioritize and address the demographic imbalance because it's more severe in this context.

```
[In]:
# Race distribution graph
df['race'].value_counts().plot(kind='bar')
[Out]:
```

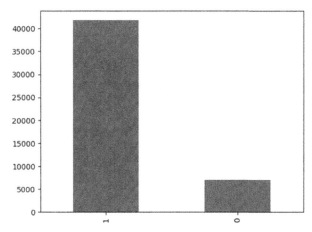

Figure 2-4. *Demographic comparison (black vs. white population)*

As shown in Figure 2-4, a class distribution of 83% to 17% is considered imbalanced in the context of machine learning. When the distribution of classes is highly imbalanced like this, a model can be biased toward the majority class, which can lead to poor performance on minority class samples.

Step 5: Model Building

Model building is a crucial step in machine learning and data analysis to make predictions or gain insights. It involves choosing the right algorithm, training the model, fine-tuning, and evaluating its performance.

- Let's build a model with biased data and see how the model performs.

- Split the dataset into train and test data and build a randomforest model on the training data as follows:

```
[In]:
# Split the data into features (x) and target (y)
x = df.drop(['income'],axis=1)
y = df['income']
# Split the data into training and testing sets
x_train,x_test, y_train, y_test = train_test_split(x,y,test_
size=0.2,random_state=42)

# Create a Random Forest Classifier model and train it
model = RandomForestClassifier(random_state=42)
model.fit(x_train,y_train)

[Out]:
RandomForestClassifier(random_state=42)
```

Step 6: Predicting for Test Data

Prediction on the test data using the model with bias dataset looks like the following:

```
[In]:
# Make predictions on the test set and calculate original
accuracy
y_pred = model.predict(x_test)
original_accuracy = accuracy_score(y_test,y_pred)
# Display the original accuracy and classification report
print('Original accuracy : ', original_accuracy)
print(classification_report(y_pred, y_test))
```

```
[Out]:
Original accuracy: 0.8525949431876344
              precision    recall  f1-score   support

           0       0.93      0.88      0.91      7814
           1       0.61      0.73      0.67      1955

    accuracy                           0.85      9769
   macro avg       0.77      0.81      0.79      9769
weighted avg       0.87      0.85      0.86      9769
```

The model performed well, with an accuracy of 85 percent. The F1 score provides a balanced measure of model performance, with higher values indicating a better balance between precision and recall. But as you can see the f1 score for 1 is 67 percent compared to 91 percent for 0 due to the bias in the data.

Let's mitigate the bias by resampling. For this, we will consider the race variable since it has 83 percent white and 17 percent black.

Step 7: Mitigating Bias

To mitigate gender bias, let's perform resampling. It specifically upsamples the minority class, which is the Black race in this example. To balance the data, we have used the resample from sklearn.

```
[In]:
# Handle class imbalance for the race column
black_candidates = df[df['race'] == 0]
white_candidates = df[df['race'] == 1]
upscaled_black = resample(black_candidates,replace=True,
n_samples=len(white_candidates), random_state=42)
balanced_data= pd.concat([upscaled_black,white_candidates])
```

Step 8: Modeling with the Balanced Data and Predicting using Test Data

Let's build a model with balanced data and see how the model performs:

```
[In]:
# Split the balanced data into features (x) and target (y)
x_balanced = balanced_data.drop(['income'],axis=1)
y_balanced = balanced_data['income']
# Split the balanced data into training and testing sets
x_train_balanced, x_test_balanced, y_train_balanced, y_test_
balanced = train_test_split(x_balanced,y_balanced,test_
size=0.2,random_state=42)
# Create a Random Forest Classifier model for balanced data and
train it
model_balanced = RandomForestClassifier(random_state=42)
model_balanced.fit(x_train_balanced, y_train_balanced)
# Make predictions on the balanced test set and calculate
balanced accuracy
y_pred_balanced = model.predict(x_test_balanced)
```

```
balanced_accuracy = accuracy_score(y_test_balanced,y_pred_
balanced)
# Display the balanced accuracy and classification report
print('Original accuracy : ', balanced_accuracy)
print(classification_report(y_pred_balanced, y_test_balanced))
```

```
[Out]:
Original accuracy: 0.9736605806644717
```

	precision	recall	f1-score	support
0	0.99	0.98	0.98	13410
1	0.92	0.95	0.93	3295
accuracy			0.97	16705
macro avg	0.95	0.96	0.96	16705
weighted avg	0.97	0.97	0.97	16705

The classification report provides insight into the performance metrics of the machine learning model for both the original and a resampled dataset.

For the original dataset, the model's accuracy is 85 percent, indicating that it correctly classifies instances 85 percent of the time. For class 0, the precision and recall are 93 percent and 88 percent, respectively, indicating that the model is correctly identifying negative instances with a high precision rate and recall rate. Conversely, for class 1, the precision and recall are 61 percent and 73 percent, respectively, suggesting that the model is less accurate at identifying positive instances. The F1 score, which is a weighted average of precision and recall, is 91 percent and 67 percent for classes 0 and 1, respectively.

For the resampled dataset, the model's accuracy is 97 percent, significantly higher than that of the original dataset. The precision and recall values for class 0 are high, with both being above 98 percent, indicating that the model is excellent at identifying negative instances. The precision and recall for class 1 are around 92 percent and 95 percent, respectively, demonstrating that the model has improved in its performance for identifying and classifying positive instances. The F1 score for classes 0 and 1 are also improved, with scores of 98 percent and 93 percent, respectively.

Overall, the resampling process has significantly improved the performance of the model, leading to improvements in accuracy, precision, recall, and F1 score.

Stage 2: Model Bias

In this activity, let's see the process of detecting bias in the model and its mitigation process. For detecting and mitigating we will be using a debiasing algorithm by AIF360 (The AI Fairness 360 is an open source python library containing techniques to help detect and mitigate bias). The goal of this algorithm is to mitigate bias in datasets and models, specifically for binary classification problems where the protected attribute is binary (e.g., gender, race, age).

The Adversarial Debiasing algorithm, part of the AI Fairness 360 toolkit, is a machine learning algorithm that aims to mitigate bias in datasets and models by training two neural networks in an adversarial manner. The algorithm has shown promise in removing bias related to protected attributes, but it is just one solution, and more work needs to be done to tackle bias in machine learning.

Dataset Details

The dataset used for data bias is also utilized here, but instead of using the raw data, we will import preprocessed data from the AIF360 package.

Source: `https://github.com/Trusted-AI/AIF360`

Step 1: Importing Packages

```
[In]:
# Import necessary libraries and modules
from aif360.datasets import BinaryLabelDataset
from aif360.datasets import AdultDataset, GermanDataset, CompasDataset
from aif360.metrics import BinaryLabelDatasetMetric
from aif360.metrics import ClassificationMetric
from aif360.metrics.utils import compute_boolean_conditioning_vector
from aif360.algorithms.preprocessing.optim_preproc_helpers.data_
preproc_functions import load_preproc_data_adult, load_preproc_
data_compas, load_preproc_data_german
from aif360.algorithms.inprocessing.adversarial_debiasing import
AdversarialDebiasing
from sklearn.linear_model import LogisticRegression
from sklearn.preprocessing import StandardScaler, MaxAbsScaler
from sklearn.metrics import accuracy_score
from IPython.display import Markdown, display
import matplotlib.pyplot as plt
import tensorflow.compat.v1 as tf
tf.disable_eager_execution()
```

Step 2: Importing the Preprocessed Dataset

For this step, instead of importing raw data and then implementing data cleaning, we use the preprocessed data from AIF360 since we covered all the preprocessing steps in data bias. See the following:

```
[In]:
# Get the dataset and split it into training and testing sets
df = load_preproc_data_adult()
privileged_groups = [{'sex': 1}]
unprivileged_groups = [{'sex': 0}]
df_train, df_test = df.split([0.7], shuffle=True)
```

Step 3: Model Building with Biased Dataset

The AdversarialDebiasing function from the AI Fairness 360 toolkit is used to define the model's scope name, privileged groups, unprivileged groups, and session. The debias parameter is set to false, indicating that the model will not be debiased:

```
[In]:
# Create a TensorFlow session
session = tf.Session()

# Create a plain model without debiasing
model = AdversarialDebiasing(privileged_groups=privileged,
                  unprivileged_groups=unprivileged,
                  scope_name='plain_classifier', debias=False,
                  sess=session)
model.fit(df_train)

[Out]:
epoch 0; iter: 0; batch classifier loss: 0.680384
epoch 0; iter: 200; batch classifier loss: 0.415313
epoch 1; iter: 0; batch classifier loss: 0.459850
epoch 1; iter: 200; batch classifier loss: 0.389641
epoch 2; iter: 0; batch classifier loss: 0.412084
epoch 2; iter: 200; batch classifier loss: 0.570294
epoch 3; iter: 0; batch classifier loss: 0.424368
epoch 3; iter: 200; batch classifier loss: 0.409188
epoch 4; iter: 0; batch classifier loss: 0.405540
epoch 4; iter: 200; batch classifier loss: 0.347155
epoch 5; iter: 0; batch classifier loss: 0.462146
epoch 5; iter: 200; batch classifier loss: 0.350587
.
.
```

```
epoch 48; iter: 0; batch classifier loss: 0.394173
epoch 48; iter: 200; batch classifier loss: 0.460599
epoch 49; iter: 0; batch classifier loss: 0.464685
epoch 49; iter: 200; batch classifier loss: 0.447618

[In]:
# Predict using the plain model
# dataset_nodebiasing_train = plain_model.predict(df_train)
y_test = model.predict(df_test)

# Close the session
session.close()
tf.reset_default_graph()
```

Step 4: Model Building with Debiased Dataset

This code initializes an adversarial debiasing model to remove bias from
the model output. The "privileged" and "unprivileged" groups are specified
to ensure that decisions are not biased against the unprivileged group:

```
[In]:
# Create a new TensorFlow session
session = tf.Session()

# Create a model with debiasing
debiased_model = AdversarialDebiasing(privileged_
groups=privileged, unprivileged_groups=unprivileged,
scope_name='debiased_classifier', debias=True, sess=session)
debiased_model.fit(df_train)

# Predict using the model with debiasing
# dataset_debiasing_train = debiased_model.predict(dataset_
orig_train)
y_debiasing_test = debiased_model.predict(df_test)
```

```
[Out]:
epoch 0; iter: 0; batch classifier loss: 0.636262; batch
adversarial loss: 0.612745
epoch 0; iter: 200; batch classifier loss: 0.486372; batch
adversarial loss: 0.658000
epoch 1; iter: 0; batch classifier loss: 0.549870; batch
adversarial loss: 0.658346
epoch 1; iter: 200; batch classifier loss: 0.472538; batch
adversarial loss: 0.644015
epoch 2; iter: 0; batch classifier loss: 0.453551; batch
adversarial loss: 0.646959
epoch 2; iter: 200; batch classifier loss: 0.445221; batch
adversarial loss: 0.619725
epoch 3; iter: 0; batch classifier loss: 0.454317; batch
adversarial loss: 0.663603
epoch 3; iter: 200; batch classifier loss: 0.386634; batch
adversarial loss: 0.583133
epoch 4; iter: 0; batch classifier loss: 0.444041; batch
adversarial loss: 0.633082
epoch 4; iter: 200; batch classifier loss: 0.490168; batch
adversarial loss: 0.597837
epoch 5; iter: 0; batch classifier loss: 0.410909; batch
adversarial loss: 0.538440
epoch 5; iter: 200; batch classifier loss: 0.443823; batch
adversarial loss: 0.642178
    .
    .
    .
epoch 47; iter: 0; batch classifier loss: 0.539765; batch
adversarial loss: 0.611614
epoch 47; iter: 200; batch classifier loss: 0.440081; batch
adversarial loss: 0.644700
```

```
epoch 48; iter: 0; batch classifier loss: 0.477121; batch
adversarial loss: 0.574099
epoch 48; iter: 200; batch classifier loss: 0.507073; batch
adversarial loss: 0.595296
epoch 49; iter: 0; batch classifier loss: 0.374277; batch
adversarial loss: 0.591830
epoch 49; iter: 200; batch classifier loss: 0.447638; batch
adversarial loss: 0.617280
```

Step 5: Comparing the Metrics of Biased and Unbiased Models

Now let's compare the results of both biased and unbiased model metrics:

```
[In]:
# Calculate classification metrics for the plain model
classified_metric_nodebiasing_test = ClassificationMetric(df_test,
                                                   y_test,
                         unprivileged_groups=unprivileged,
                         privileged_groups=privileged)

# Display classification metrics for the plain model
display(Markdown("Model without debiasing metrics"))
print("Accuracy = %f" % classified_metric_nodebiasing_test.
accuracy())
print("Disparate impact = %f" % classified_metric_nodebiasing_
test.disparate_impact())
print("Equal opportunity difference = %f" % classified_metric_
nodebiasing_test.equal_opportunity_difference())
print("Average odds difference = %f" % classified_metric_
nodebiasing_test.average_odds_difference())
```

```
# Calculate classification metrics for the model with debiasing
classified_metric_debiasing_test = ClassificationMetric(df_test,
                                                y_debiasing_test,
                                    unprivileged_groups=unprivileged,
                                        privileged_groups=privileged)

# Display classification metrics for the model with debiasing
display(Markdown("Model with debiasing metrics"))
print("Accuracy = %f" % classified_metric_debiasing_test.
accuracy())
print("Disparate impact = %f" % classified_metric_debiasing_
test.disparate_impact())
print("Equal opportunity difference = %f" % classified_metric_
debiasing_test.equal_opportunity_difference())
print("Average odds difference = %f" % classified_metric_
debiasing_test.average_odds_difference())

[Out]:
Model without debiasing metrics
Accuracy = 0.809254
Disparate impact = 0.000000
Equal opportunity difference = -0.460858
Average odds difference = -0.282582

Model with debiasing metrics
Accuracy = 0.796014
Disparate impact = 0.561960
Equal opportunity difference = -0.078535
Average odds difference = -0.046530
```

The output consists of two models: one without debiasing metrics and the other with debiasing metrics.

For the model without debiasing metrics, the accuracy is 0.809254. The `disparate impact` metric is 0, indicating that there is no difference in the proportion of positive classifications between the protected and unprotected groups. The `equal opportunity difference` metric is −0.460858, which means that the difference in true positive rates between protected and unprotected groups is significant. Finally, the `average odds difference` metric is −0.282582, which means that there is a difference in both true and false positive rates between protected and unprotected groups.

For the model with debiasing metrics, the accuracy is slightly lower at 0.796014. The `disparate impact` metric is 0.561960, indicating that there is a difference in positive classifications between the protected and unprotected groups, but it is less severe than in the previous model. The `equal opportunity difference` metric is −0.078535, indicating that the difference in true positive rates between protected and unprotected groups is less severe than in the previous model. Finally, the `average odds difference` metric is −0.046530, indicating that there is a smaller difference in true and false positive rates between protected and unprotected groups than in the previous model.

Overall, the model with debiasing metrics shows improvement in mitigating bias compared to the model without debiasing metrics, as the `disparate impact`, `equal opportunity difference`, and `average odds difference` metrics are all lower. However, it is important to note that the accuracy of the model has slightly.

Conclusion

In this chapter, our exploration unveiled how bias influences decisions, from individual judgments to complex systems. Recognizing bias's origins and impact is crucial for cultivating equitable AI.

This journey uncovers various biases arising from cognitive and societal factors with the potential to perpetuate inequality. However, through tools like fairness-aware training and explainable AI, we can address bias and promote transparent, just systems.

Ultimately, this quest equips us to navigate bias and fairness complexities, shaping AI that respects values and embraces diversity. As AI evolves, our understanding guides us toward an inclusive future, where fairness remains the guiding principle.

## CHAPTER 3

Transparency and Explainability

In the rapidly advancing landscape of artificial intelligence (AI), the decisions made by AI systems are playing an increasingly pivotal role in various aspects of our lives. As AI technologies grow progressively more sophisticated and seamlessly integrate into pivotal domains such as health care, finance, judicial proceedings, and autonomous systems, it becomes an urgent necessity to gain a comprehensive understanding of the mechanisms underpinning these systems' decision-making processes. This underscores the paramount importance of embracing the principles of transparency and explicability.

Transparency

"Transparency" refers to the openness and clarity with which an AI system's decision-making process can be understood. Essentially, it involves shedding light on the internal mechanics, algorithms, and data used by an AI system to reach conclusions. Transparent systems enable stakeholders, including developers, users, regulators, and even the general public, to understand why a particular decision was made.

Explainability

Explainability goes a step further by not only revealing the decision-making process but also providing understandable explanations for those decisions. An explainable AI system can describe the factors, features, or data points that influence a specific output. This not only builds trust but also helps in diagnosing errors, identifying biases, and addressing potential ethical concerns within the AI's decision-making process.

Importance of Transparency and Explainability in AI Models

Transparency and explainability are fundamental to ensuring that AI systems are trustworthy, accountable, and aligned with ethical principles. Here's how these concepts contribute to these important aspects:

1. **Trust:**

   Transparency and explainability are cornerstones of building trust in AI systems. When users, stakeholders, and the general public can understand how an AI system reaches its conclusions, they are more likely to trust its outputs. Trust is especially crucial in domains where AI decisions impact critical aspects of human life, such as health-care diagnoses, legal judgments, or financial transactions. Transparent and explainable AI reassures individuals that the technology is not making arbitrary or biased decisions.

2. **Accountability:**

   Accountability is deeply linked to transparency. When AI systems are transparent, developers and organizations can be held accountable for the functioning and outcomes of their technology. If an AI system makes a questionable or incorrect decision, being transparent about the decision-making process enables organizations to investigate and rectify the issue. This also extends to regulatory compliance, as transparent systems are easier to audit and ensure adherence to ethical and legal standards.

3. **Ethical Considerations**

   AI systems can unintentionally amplify biases present in their training data, giving rise to inequitable or prejudiced results. The principles of transparency and explainability play a pivotal role in addressing this ethical concern. When AI systems offer justifications for their decisions, it facilitates the identification of any biased patterns impacting those decisions. This empowers developers and researchers to actively mitigate biases and ensure the equitable and impartial behavior of AI systems.

4. **Informed Decision Making**

   Transparent and explainable AI empowers users and stakeholders to make informed decisions based on AI-generated insights. For instance, in the medical field, doctors need to understand why an AI system arrived at a certain diagnosis to determine its accuracy and relevance. Similarly, consumers making financial decisions based on AI recommendations can better assess the validity and rationale behind those recommendations.

5. **Error Detection and Improvement:**

   Transparent and explainable AI systems facilitate the detection of errors, anomalies, or unexpected behavior. When the decision-making process is clear, deviations from expected outcomes can be swiftly identified and corrected. This proactive approach enhances the reliability and robustness of AI systems over time.

In the field of AI, transparency and explainability are foundational principles that cultivate trust, accountability, and the ethical development and implementation of AI. These ideas serve to connect the intricate internal mechanisms of AI systems with the human necessity for understanding and rationale. As AI assumes an increasingly substantial role in society, having transparent and explainable AI systems becomes paramount in crafting technology that serves the betterment of humanity while maintaining ethical benchmarks.

Real-world Examples of the Impact of Transparent AI

Transparent AI can have a significant impact on various real-world scenarios, such as

1. **Health-care Diagnosis:**

   Transparent AI can significantly impact health care by enhancing diagnostic accuracy and transparency. For instance, IBM's Watson for Oncology uses transparent AI to assist doctors in cancer diagnosis and treatment recommendations. The system provides explanations for its recommendations, allowing doctors to understand the reasoning behind each suggestion. This transparency improves doctor–patient communication and builds trust in the treatment process.

2. **Credit Scoring:**

   In the financial sector, transparent AI can address biases in credit scoring. Upstart, a lending platform, uses transparent AI to assess creditworthiness. Traditional credit-scoring methods sometimes discriminate against individuals with limited credit histories. Upstart's transparent AI incorporates additional factors, such as education and job history, providing explanations for loan decisions and promoting fairness.

3. **Criminal Justice:**

   Transparent AI can mitigate bias in criminal justice systems. Northpointe's COMPAS algorithm, used to assess recidivism risk, faced criticism for potential racial bias. ProPublica's investigation revealed disparities in risk assessments. Transparent AI solutions, like the one developed by the Stanford Computational Policy Lab, aim to provide clear explanations for risk predictions, allowing judges and defendants to understand and question the outcomes.

4. **Autonomous Vehicles:**

   Transparent AI is crucial for building trust in autonomous vehicles. Waymo, a subsidiary of Alphabet Inc., is a pioneer in this field. Their self-driving cars provide real-time explanations for driving decisions, such as lane changes and braking. Transparent AI helps passengers understand the vehicle's actions and builds confidence in its ability to navigate complex traffic scenarios.

5. **Natural Language Processing:**

Transparent AI is transforming natural language processing (NLP) applications. OpenAI's GPT-3, a highly advanced language model, can generate human-like text. To ensure transparency, OpenAI provides guidelines for users to understand how to interact responsibly with the technology, reducing the risk of biased or inappropriate outputs.

6. **Medicine and Drug Discovery:**

Transparent AI aids in drug discovery and medical research. BenevolentAI uses AI to identify potential drug candidates for various diseases. Their platform explains why certain molecules are selected as drug candidates based on their properties and interactions. This transparency assists researchers in making informed decisions about which molecules to prioritize.

7. **Online Content Moderation:**

Transparent AI is valuable in content moderation on platforms like YouTube. Google's Perspective API employs transparent AI to flag and evaluate potentially inappropriate content in comments and discussions. By providing insights into the reasons for content moderation decisions, users gain a better understanding of community guidelines.

8. **Environmental Monitoring:**

Transparent AI contributes to environmental monitoring and conservation efforts. For instance, Wildlife Insights utilizes AI to identify and track wildlife in camera-trap images. The system provides transparent classifications, allowing researchers to validate and correct identification errors, ensuring accurate wildlife population data.

Transparent AI is reshaping various sectors by fostering accountability, fairness, and user trust. From health care and finance to criminal justice and autonomous vehicles, real-world examples showcase how transparent AI enhances decision-making processes, promotes fairness, mitigates biases, and ultimately contributes to the responsible and ethical use of AI technology.

Methods for Achieving Explainable AI

Achieving Explainable AI (XAI) is a critical endeavor in the field of artificial intelligence, driven by the need to make complex machine learning models more transparent and interpretable. Some methods for achieving XAI include:

Explanation Methods for Interpretable Models: Decision Trees and Rule-Based Systems

Interpretable models are essential for domains where transparency and understanding of decision making are critical. Decision trees and rule-based systems are examples of models that provide clear explanations for their predictions. The explanation methods for these models play a pivotal role in making AI understandable and trustworthy:

1. **Decision Trees:** A decision tree is a hierarchical structure that segments data by asking a series of questions. Each decision in the tree narrows down the possibilities until a final prediction is reached.

- **Feature Importance:** Decision trees assign importance scores to features based on how much they contribute to the model's decisions. Features with higher importance influence predictions more.

- **Path Explanation:** By tracing the path from the root to a specific leaf node, one can understand how a decision tree arrives at a particular prediction. The sequence of decisions serves as a rationale.

- **Visual Representation:** Visualization tools depict decision trees graphically. Users can follow the branches and nodes to grasp the decision process intuitively.

- **Rules Extraction:** Decision trees can be transformed into a set of if–then rules. These rules outline the conditions that lead to specific predictions, offering explicit insights.

2. **Rule-Based Systems:** Rule-based systems operate on predefined rules that dictate how inputs are mapped to outputs.

    - **Human-Readable Rules:** Rules in rule-based systems are formulated in natural language or a simplified syntax, making them easily understandable even by non-technical users.

    - **Rule Weights and Certainties:** Some rule-based systems assign weights or certainties to rules, indicating their significance. This provides insight into the relative importance of rules.

- **Rule Set Simplification:** Complex rule sets can be simplified by removing redundant or less impactful rules. This streamlines the explanation process.

3. **Comparative Explanation:** Rule-based systems can be compared against human expertise or regulatory guidelines. When rules align with human knowledge, it boosts trust.

4. **Hybrid Approaches:** Combining the strengths of decision trees and rule-based systems can result in more-comprehensive explanations. Decision tree rules can be extracted and integrated into a rule-based system.

Explanation methods for interpretable models like decision trees and rule-based systems are instrumental in making AI transparent and trustworthy. By providing interpretable rationales for predictions, these methods bridge the gap between AI's complexity and human understanding. In domains where accountability, fairness, and human collaboration are paramount, these methods empower stakeholders to make informed decisions and ensure responsible AI use across diverse applications.

Generating Feature Importance Scores and Local Explanations

Interpreting the decision-making process of machine learning models is crucial for understanding their behavior. Generating feature importance scores and local explanations are techniques used to unravel the contributions of individual features and provide insights into model predictions on a per-instance basis (Figure 3-1).

1. **Generating Feature Importance Scores:**

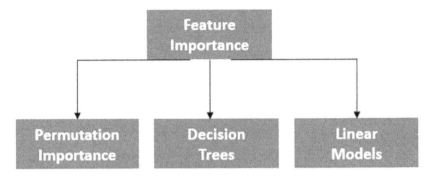

Figure 3-1. *Types of feature importance methods*

- **Overview:** Feature importance scores quantify the influence of each input feature on a model's predictions. These scores help determine which features are most relevant in driving model outcomes.

    - **Permutation Importance:** Permutation importance involves shuffling the values of a single feature while keeping other features unchanged. The decrease in model performance (e.g., accuracy) when that feature is permuted reflects its importance. Features causing significant performance drops are considered more important.

    - **Mean Decrease Impurity (Decision Trees):** For decision tree–based models, feature importance can be measured using mean decrease impurity. This metric calculates the reduction in impurity (e.g., Gini impurity) due to splits using a particular feature. Larger reductions imply greater feature importance.

- **Coefficient Magnitudes (Linear Models):** In linear models, feature importance can be inferred from the magnitudes of the coefficients. Larger coefficients indicate stronger feature influences on the output.

2. **Generating Local Explanations:**

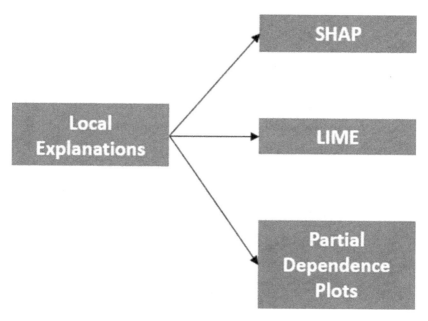

Figure 3-2. *Types of local explanation*

- **Overview:** Local explanations focus on interpreting model predictions for individual instances rather than the entire dataset (Figure 3-2). They help users understand why a model produced a specific output for a given input.

- **LIME (Local Interpretable Model-agnostic Explanations):** LIME generates a simplified surrogate model around the instance of interest. Perturbed samples are created by altering the input features, and the surrogate model learns to mimic the predictions of the black-box model. The surrogate model's explanation, like feature importance scores, highlights influential features.

- **SHAP (SHapley Additive exPlanations):** SHAP attributes each prediction's deviation from the average model output to specific features. It considers all possible feature combinations and calculates their contributions. Positive SHAP values indicate features pushing the prediction higher, while negative values indicate features pulling the prediction lower.

- **Partial Dependence Plots (PDP):** PDP illustrates how the variation in a single feature affects model predictions while keeping other features constant. It helps visualize the relationship between a feature and the model's output.

3. **Creating Saliency Maps:** Interpreting the decisions of machine learning models is essential for building trust and ensuring the ethical use of AI. Saliency maps are visualization tools that help us understand how a model's predictions are influenced by different features in the input data. Saliency maps highlight the most relevant areas of an input image or data instance that contribute to the model's decision-making process.

- **Overview:** Saliency maps are generated by analyzing the gradients of the model's output with respect to the input features. Gradients indicate how sensitive the model's predictions are to changes in input features.

  - **Gradient-Based Methods:** Gradient-based methods, such as Gradient-weighted Class Activation Mapping (Grad-CAM), compute the gradients of the model's predicted class score with respect to the feature map of the final convolutional layer. These gradients are used to weight the feature maps, highlighting regions that strongly influence the prediction.

  - **Backpropagation:** Backpropagation calculates the gradients of the model's output with respect to the input features. These gradients indicate how much each input feature contributes to the final prediction.

Generating feature importance scores and local explanations are essential techniques for interpreting machine learning models. They shed light on which features matter most and help users understand why a model behaves the way it does for individual instances. These techniques foster trust, accountability, and transparency, contributing to responsible AI deployment across various domains.

Saliency maps are invaluable tools for understanding the inner workings of machine learning models. By visualizing the input features that drive a model's decisions, saliency maps promote transparency, trust, and accountability. These maps empower users, experts, and regulators to comprehend model behavior, identify biases, and ensure ethical and responsible AI deployment across diverse applications.

Tools, Frameworks, and Implementation of Transparency and Explainability

Tools, frameworks, and the implementation of transparency and explainability in AI are instrumental in making complex machine learning models comprehensible and accountable. These resources encompass a variety of software libraries, methodologies, and practices that enable data scientists and developers to achieve greater transparency and interpretability in AI systems. By utilizing these tools and frameworks, organizations and researchers can enhance the trustworthiness and ethical use of AI in real-world applications.

Overview of Tools and Libraries for AI Model Transparency

As the field of artificial intelligence advances, having transparency and explainability in AI models becomes increasingly important. Several tools and libraries have been developed to help practitioners, researchers, and organizations understand, interpret, and make responsible use of AI models. These tools assist in visualizing, explaining, and evaluating the behavior of AI models.

1. **InterpretML:** InterpretML is an open-source library that provides various techniques for interpreting machine learning models. It supports black-box models and includes methods like SHAP (SHapley Additive exPlanations) and LIME (Local Interpretable Model-agnostic Explanations).

   Features:

   - Supports various machine learning models, both black-box models refer to complex algorithms, such as deep neural networks, which are challenging to interpret due to their intricate internal workings and white-box models, on the other hand, are simpler and more interpretable, like decision trees or linear regression.

- Implements SHAP (SHapley Additive exPlanations) and LIME (Local Interpretable Model-agnostic Explanations) methods.

- Provides global and local explanations for model predictions.

- Suitable for tabular, text, and image data.

- Offers user-friendly APIs for easy integration and usage.

2. **SHAP (SHapley Additive exPlanations):** SHAP is a framework for explaining the output of any machine learning model. It is based on cooperative game theory and calculates the contribution of each feature to a model's prediction.

Features:

- Is a versatile framework based on cooperative game theory.

- Computes feature importance values for explaining model predictions.

- Offers Shapley values, which ensure fair distribution of credit among features.

- Provides unified explanations for various model types, both global and local.

- Enables quantification of feature contributions.

3. **LIME (Local Interpretable Model-agnostic Explanations):** LIME is a technique used to explain individual predictions by training interpretable surrogate models around specific instances. It perturbs input data and observes the impact on model predictions.

Features:

- Focuses on explaining individual predictions made by black-box models.

- Generates locally interpretable surrogate models to approximate the behavior of the black-box model.

- Uses perturbed samples to train a simple interpretable model around a specific instance.

- Offers both tabular and text data support for generating explanations.

- Provides insights into feature importance for specific predictions.

- Suitable for understanding complex model behavior at the instance level.

4. **TensorBoard:** TensorBoard is a visualization toolkit developed by Google for understanding and debugging machine learning models built with TensorFlow. It provides visualizations of training metrics, model graphs, and embeddings.

Features:

- Developed by Google, TensorBoard is primarily a visualization toolkit for TensorFlow models.

- Offers visualizations for training metrics, model graphs, histograms, embeddings, and more.

- Helps users track model training progress, understand complex architectures, and diagnose issues.

- Enables visualization of model behavior, weights, and layer activations.

- Integrates seamlessly with TensorFlow-based projects.

- Useful for analyzing model performance, debugging, and model interpretation through visual insights.

5. **XAI (Explainable AI) Libraries:** Various XAI libraries, such as Alibi, provide a range of techniques to explain and interpret AI models. Alibi, for instance, includes methods like counterfactual explanations and anchors.

Features:

- Offers various interpretable machine learning methods.

- Provides LIME and SHAP explanations for model predictions.

- Supports multiple model types, including regression and classification.

- Includes visualization tools for better understanding.

- Is useful for generating local and global explanations.

6. **ELI5 (Explain Like I'm 5):** ELI5 is a Python library that provides simple explanations of machine learning models. It supports multiple model types and includes methods like feature importance and permutation importance.

   Features:

   - Simplifies explanations of machine learning models.

   - Supports various model types, including tree-based, linear, and text models.

   - Provides feature importances, permutation importance, and coefficients.

   - Offers straightforward explanations for non-technical users.

   - Suitable for quick model assessment and interpretation.

7. **Yellowbrick:** Yellowbrick is a Python library for visualizing machine learning model behavior. It offers tools for assessing model performance, selecting features, and diagnosing common issues.

   Features:

   - Focuses on visualization of model behavior and performance.

   - Provides visual tools for assessing model training, selection, and debugging.

   - Includes various visualizers for ROC curves, feature importances, residuals, etc.

- Is compatible with Scikit-Learn and other popular libraries.

- Aids in understanding complex model outputs through visual insights.

Tools and libraries designed to facilitate AI model transparency play a crucial role in making machine learning models more understandable, accountable, and ethically sound. These resources empower users to uncover insights, explain model decisions, identify biases, and ensure responsible AI deployment across various domains. As the field evolves, these tools contribute to building trust between AI systems and their users.

Implementation of Explainable AI

Incorporating explainable AI provides numerous advantages. For decisionmakers and involved parties, it delivers a thorough understanding of the logic underlying AI-driven decisions, facilitating the enhancement of well-informed selections. Additionally, it aids in pinpointing potential biases or inaccuracies in the models, culminating in outcomes that are both more precise and equitable.

About Dataset

The data were gathered and provided by the National Institute of Diabetes and Digestive and Kidney Diseases as a component of the Pima Indians Diabetes Database. Specific criteria were applied when choosing these instances from a broader database. Notably, all the individuals featured here are of Pima Indian ancestry (a subset of Native Americans) and are females aged 21 and older.

Source: `https://www.kaggle.com/datasets/kandij/diabetes-dataset?resource=download`

Let's take a look at the dataset's features:

- Pregnancies: Number of times pregnant

- Glucose: Plasma glucose concentration at two hours in an oral glucose tolerance test

- BloodPressure: Diastolic blood pressure (mm Hg)

- SkinThickness: Triceps skin fold thickness (mm)

- Insulin: 2-Hour serum insulin (mu U/ml)

- BMI: Body mass index (weight in kg/(height in m)^2)

- DiabetesPedigreeFunction: Diabetes pedigree function

- Age: Age (years)

- Outcome: Diastolic/Non-diastolic (Target)

Getting Started

The goal is to implement an AI solution that exemplifies the principles of transparency, explainability, and ethicality. The objective is to strike a balance between model performance and interpretability, leveraging techniques such as model-agnostic explanations, feature importance analysis, and visualization methods. The implemented solution should address real-world challenges and promote user trust.

Stage 1: Model Building

Model building is a crucial step in machine learning and data analysis to make predictions or gain insights. It involves choosing the right algorithm, training the model, fine-tuning, and evaluating its performance.

Step 1: Import the Required Libraries

```
[In]:
# Import necessary libraries
import pandas as pd
import seaborn as sns
import matplotlib.pyplot as plt
from sklearn.model_selection import train_test_split
from sklearn.tree import DecisionTreeClassifier, plot_tree
from sklearn.ensemble import RandomForestClassifier
from sklearn.metrics import classification_report

import shap
from lime.lime_tabular import LimeTabularExplainer
import eli5
```

Step 2: Load the Diabetes Dataset

```
[In]:
# Read the dataset into a pandas DataFrame
df = pd.read_csv('diabetes.csv')
df.head()
```

```
[Out]:
```

Pregnancies	Glucose	Blood Pressure	Skin Thickness	Insulin	BMI	Diabetes Pedigree Function	Age	Outcome
6	148	72	35	0	33.6	0.627	50	1
1	85	66	29	0	26.6	0.351	31	0
8	183	64	0	0	23.3	0.672	32	1
1	89	66	23	94	28.1	0.167	21	0

Step 3: Checking the Data Characteristics

Check if there are any discrepancies, like missing values, wrong data types, etc.

```
[In]:
# Display basic information about the dataset
df.info()
```

```
[Out]:
<class 'pandas.core.frame.DataFrame'>
RangeIndex: 768 entries, 0 to 767
Data columns (total 9 columns):
 #   Column                    Non-Null Count   Dtype
---  ------                    --------------   -----
 0   Pregnancies               768 non-null     int64
 1   Glucose                   768 non-null     int64
 2   BloodPressure             768 non-null     int64
 3   SkinThickness             768 non-null     int64
 4   Insulin                   768 non-null     int64
 5   BMI                       768 non-null     float64
 6   DiabetesPedigreeFunction  768 non-null     float64
 7   Age                       768 non-null     int64
 8   Outcome                   768 non-null     int64
dtypes: float64(2), int64(7)
memory usage: 54.1 KB
```

There are no null values present in the data, so we can proceed with the data preprocessing steps.

Step 4: Exploratory Data Analysis

This is a summary of statistical measures for each column.

```
[In]:
df.describe()
```

```
[Out]:
```

	Pregnancies	Glucose	Blood Pressure	Skin Thickness	Insulin	BMI	Diabetes Pedigree Function	Age	Outcome
count	768.000000	768.000000	768.000000	768.000000	768.000000	768.000000	768.000000	768.000000	768.000000
mean	3.845052	120.894531	69.105469	20.536458	79.799479	31.992578	0.471876	33.240885	0.348958
std	3.369578	31.972618	19.355807	15.952218	115.244002	7.884160	0.331329	11.760232	0.476951
min	0.000000	0.000000	0.000000	0.000000	0.000000	0.000000	0.078000	21.000000	0.000000
25%	1.000000	99.000000	62.000000	0.000000	0.000000	27.300000	0.243750	24.000000	0.000000
50%	3.000000	117.000000	72.000000	23.000000	30.500000	32.000000	0.372500	29.000000	0.000000
75%	6.000000	140.250000	80.000000	32.000000	127.250000	36.600000	0.626250	41.000000	1.000000
max	17.000000	199.000000	122.000000	99.000000	846.000000	67.100000	2.420000	81.000000	1.000000

```
[In]:
sns.countplot(x='Outcome',data=df)

[Out]:
```

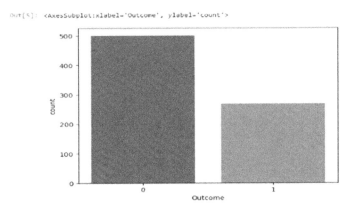

Figure 3-3. *Distribution of the outcome*

A 37 percent to 63 percent class distribution is slightly imbalanced, but whether it's acceptable depends on your specific problem and algorithm. Some algorithms can handle this, while others may require balancing techniques like oversampling or undersampling.

Step 5: Model Building

Now we split the dataset for training and testing. A random forest classifier is built to predict diabetes outcomes using the diabetes dataset.

```
[In]:
# Separate train and test Variables
x = df.drop('Outcome',axis=1)
y = df['Outcome']

# Create Train & Test Data
```

```
x_train,x_test,y_train,y_test = train_test_split(x,y,test_
size=0.3,random_state=101)

# Build the model
model = RandomForestClassifier(max_features=2, n_estimators =100,
bootstrap = True)
model.fit(x_train, y_train)
```

Step 6: Predicting for Test Data

```
[In]:
# Predict the testing data and build a classification report
y_pred = model.predict(x_test)
print(classification_report(y_pred, y_test))
```

[Out]:

	precision	recall	f1-score	support
0	0.83	0.81	0.82	155
1	0.63	0.67	0.65	76
accuracy			0.76	231
macro avg	0.73	0.74	0.73	231
weighted avg	0.77	0.76	0.76	231

The random forest classifier gives a good performance in predicting diabetes outcomes, with room for improvement, but for now we will continue with this.

Now, let's integrate the explainability layer into the model to provide more insight into the output. The next section will focus on the two categories of model explainability: model-specific methods and model-agnostic methods.

Stage 2: SHAP

Model-agnostic methods can be applied to any machine learning model, regardless of its type. They focus on analyzing the feature's input and output pair. This section will introduce and discuss SHAP, the widely used models.

Step 1: Creating an Explainer and Feature Importance Plot

```
[In]:
# Create the explainer
explainer = shap.TreeExplainer(model)
shap_values = explainer.shap_values(x_test)
# Variable Importance Plot
figure = plt.figure()
shap.summary_plot(shap_values, x_test)
[Out]:
```

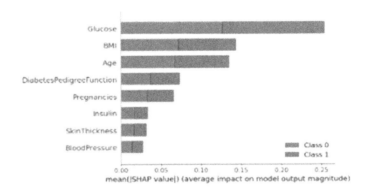

Figure 3-4. *SHAP summary plot*

Using the `summary_plot()` function, the features are sorted according to their average SHAP values, with the most significant features at the top and the least significant ones at the bottom. This makes it easier to comprehend how each feature affects the predictions made by the model.

The plot in Figure 3-4 shows that each class is equally occupied in the first three bars. This indicates that each characteristic affects the prediction of both diabetes (1) and non-diabetes (0) in an equal manner. It demonstrates that the first three features have the best ability to forecast. However, pregnancy, skin thickness, insulin, and blood pressure do not contribute as significantly.

Step 2: Summary Plot

```
[In]:
shap.summary_plot(shap_values[1], x_test)
[Out]:
```

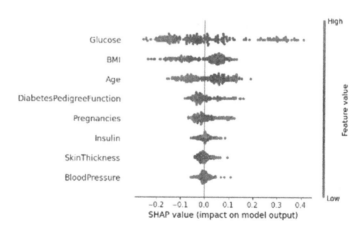

Figure 3-5. *SHAP output impact plot*

This method can give a more detailed perspective of how each attribute affects a particular result.

- In the plot in Figure 3-5, positive values push the model's prediction closer to diabetes. In contrast, negative values push toward nondiabetes.

- Elevated glucose levels may be associated with an increased risk of diabetes, while low glucose levels are not indicative of the absence of diabetes.

- Similarly, aged patients are more likely to be diagnosed with diabetes. However, the model seems uncertain about the diagnosis for younger patients.

The way of dealing with this for the Age attribute is by using the dependence plot to get more-detailed insights.

Step 3: Dependence Plot

Dependence plots display the relationship between a feature and the outcome for each occurrence of the data. It goes beyond just gathering additional specific data and verifying the significance of the feature under analysis by confirming or challenging the results of other global feature importance measures.

```
[In]:
shap.dependence_plot('Age', shap_values[1], x_test,
interaction_index="Age")
[Out]:
```

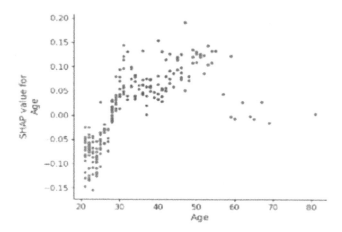

Figure 3-6. *SHAP dependence plot*

The dependence plot reveals that patients under age 30 have a lower risk of being diagnosed with diabetes. In contrast, individuals over 30 face a higher likelihood of receiving a diabetes diagnosis.

Stage 3: LIME

Let's fit the LIME explainer using the training data and the targets. Set the mode to classification.

Step 1: Fitting the LIME Explainer

```
[In]:
class_names = ['Diabetes', 'No diabetes']
feature_names = list(x_train.columns)
explainer = LimeTabularExplainer(x_train.values,
feature_names =
                          feature_names,
                          class_names = class_names,
                          mode = 'classification')
```

Step 2: Plotting the Explainer

Let's display a LIME explanation for the first row in the test data using the random forest, and show the feature contribution.

The result contains three main pieces of information:

1. The model's predictions

2. Features' contributions

3. The actual value for each feature

```
[In]:
explainer = explainer.explain_instance(x_test.iloc[1], model.
predict_proba)
explainer.show_in_notebook(show_table = True, show_all = False)
[Out]:
```

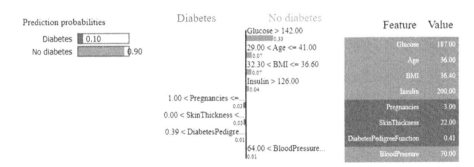

Figure 3-7. *LIME explainer*

We observe that it predicted 90 percent as non-diabetic because, as we see in Figure 3-7,

- the glucose level is greater than 142;

- age is between 29 and 41;

- BMI is between 32 and 36; and

- the insulin is greater than 126.

Stage 4: ELI5

ELI5 is a Python package that allows the inspection of machine learning classifiers and provides explanations for their predictions. It is a commonly used tool for debugging machine learning algorithms, such as scikit-learn regressors and classifiers, XGBoost, CatBoost, Keras, etc. With ELI5, developers and data scientists can better understand how a model arrived at a certain result, identify any limitations or issues with the model, and ultimately improve the accuracy and reliability of their models.

Source: https://github.com/TeamHG-Memex/eli5

Step 1: Viewing Weights for the Fitted Model

This displays the weights of a model and their corresponding feature names using ELI5.

```
[In]:
# Get a list of all feature names/columns in the dataset
all_features = x.columns.to_list()
# Use ELI5 to show the weights of the model, with feature names
provided
eli5.show_weights(model, feature_names=all_features)

[Out]:
```

Weight	Feature
0.2678 ± 0.1203	Glucose
0.1609 ± 0.1013	BMI
0.1385 ± 0.0899	Age

Weight	Feature
0.1244 ± 0.0692	DiabetesPedigreeFunction
0.0895 ± 0.0565	BloodPressure
0.0869 ± 0.0635	Pregnancies
0.0670 ± 0.050	SkinThickness
0.0652 ± 0.0639	Insulin

The preceding table gives us the weight associated with each feature. The value tells us how much of an impact a feature has on the predictions on average, and the sign tells us in which direction. Here, we can observe that the "Glucose" feature has the highest weight of 0.2678 ± 0.1203, implying that it has the most significant impact on the algorithm's predictions.

This table can be used to identify the most impactful features that are highly relevant to predictions made by an algorithm.

Step 2: Explaining for the Test Data

This displays ELI5's prediction interpretation with feature values.

```
[In]:
# Use ELI5 to show the prediction for one of the test samples
(again, in this case, sample 1)
# The show_feature_values parameter is set to True to display
feature values
eli5.show_prediction(model, x_test.iloc[1], feature_names=all_
features, show_feature_values=True)

[Out]:
y=1 (probability 0.840) top features
```

Contribution	Feature	Value
0.348	<BIAS>	1
0.325	Glucose	187
0.083	BMI	36.4
0.047	Age	36
0.019	DiabetesPedigreeFunction	0.408
0.011	BloodPressure	70
0.01	Insulin	200
0.001	Pregnancies	3
-0.005	SkinThickness	22

This table denotes that the given instance was classified as "1" with a probability of 0.840. The table shows the top features that contributed most to this prediction in descending order, based on the values of the weights assigned to each feature when the prediction was made.

The first row of the table with a value of 0.348 represents the bias term, which is the default value added to each prediction and has an inherent contribution to the model output. The contribution of other features is compared to this bias term. We can observe that the feature "Glucose" has the highest contribution of 0.325, followed by "BMI" with a contribution of 0.083 and "Age" with a contribution of 0.047. The "DiabetesPedigreeFunction," "BloodPressure," "Insulin," "Pregnancies," and "SkinThickness" features were less impactful, with smaller contributions ranging from 0.001 to 0.019.

This output aids in understanding the relative importance of various features in the prediction of the outcome by providing a rough idea of how much each feature contributed to the final prediction of that instance.

Stage 5: Conclusion

This exercise has offered a decent introduction to what explainable AI is and some principles that contribute to developing trust and that can provide data scientists and other stakeholders with the necessary skill sets to assist in making meaningful decisions.

Challenges and Solutions in Achieving Transparency and Explainability

Complexity challenges in deep learning and complex AI architectures are significant hurdles that arise due to the intricate nature of advanced models and the increasing demands for higher performance. These challenges encompass various aspects, including model interpretability, training difficulties, computational resources, and ethical considerations. Let's delve into these challenges in detail:

1. **Black-Box Nature:** Deep learning models, particularly neural networks, consist of multiple layers with numerous parameters. This complexity enables them to capture intricate patterns in data.

   Challenge: Highly complex models can be challenging to interpret and understand. As the number of layers and parameters increases, the decision-making process becomes more opaque.

2. **Data-Driven Complexity:** Complex AI architectures, including deep learning models, are data-hungry and require massive amounts of data for training.

   Challenge: Acquiring and preprocessing large datasets can be resource-intensive and time-consuming. Biases present in data can also lead to biased model outputs.

3. **Accuracy and Interpretability:** The trade-off between accuracy and interpretability in AI revolves around the fact that as models become more accurate and complex, their decision-making process becomes harder to understand. While highly accurate models often involve intricate architectures that can achieve remarkable results, their complexity can obscure the reasons behind their predictions. However, more interpretable models tend to be simpler and easier to understand, but they might sacrifice some accuracy. Striking the right balance requires careful consideration of the application's requirements, ethical concerns, and the need for transparency to build trust with users and stakeholders.

   Challenge: Balancing the trade-off between accuracy and interpretability presents challenges in determining the ideal equilibrium where model complexity guarantees both high accuracy and comprehensible outcomes. Complex models with high accuracy can pose difficulties in interpretation, raising issues of reduced user confidence and regulatory implications. Conversely, models designed for extreme interpretability might compromise accuracy, constraining their real-world applicability. Achieving equilibrium involves managing ethical aspects, adhering to regulations, and mitigating potential biases, all while providing precise outcomes that users can readily understand and rely upon.

Addressing the "Black Box" Nature of AI Models

Addressing the "black box" nature of AI models involves efforts to make complex machine learning algorithms more transparent and understandable. This is crucial for ensuring that AI systems can be trusted, validated, and their decision-making processes can be explained and interpreted. Techniques, tools, and methodologies are employed to shed light on how AI models arrive at their predictions and why, enhancing accountability, fairness, and ethical use in various applications.

1. **Model Selection:**

   Explanation: Choose models with inherent interpretability or that produce understandable outputs.

   Approach: Opt for models like decision trees, linear regression, or rule-based systems, which are more interpretable. These models provide insights into how features influence predictions.

2. **Simplification of Architecture:**

   Explanation: Use simpler versions of complex models for easier interpretability.

   Approach: Instead of deploying a deep neural network, consider using a smaller architecture with fewer layers, making it more understandable.

3. **Hybrid Models:**

   Explanation: Combine complex models with interpretable components.

   Approach: Build hybrid models that include interpretable components like decision rules, allowing users to understand certain parts of the decision process.

4. **Local Explanations:**

   Explanation: Provide explanations for individual predictions.

   Approach: Implement techniques like LIME (Local Interpretable Model-agnostic Explanations) or SHAP (SHapley Additive exPlanations) to generate explanations for specific predictions.

5. **Feature Importance Analysis:**

   Explanation: Understand the contribution of features to predictions.

   Approach: Use techniques like SHAP or feature importance scores to identify which features have the most impact on model outputs.

6. **Surrogate Models:**

   Explanation: Create simplified surrogate models that mimic the behavior of the black-box model.

   Approach: Train interpretable models, such as linear regression, to approximate the complex model's predictions. These surrogate models offer insights into the decision process.

7. **Visualization Techniques:**

   Explanation: Visualize model behavior and predictions.

   Approach: Utilize visualization tools like SHAP plots, saliency maps, or feature contribution graphs to make the model's decision-making process more understandable.

8. **Counterfactual Explanations:**

   Explanation: Generate examples of input changes needed for a different model outcome.

   Approach: Explain model predictions by showing users which modifications to input data would lead to a different prediction.

9. **Contextual Information:**

   Explanation: Incorporate domain-specific knowledge to contextualize model outputs.

   Approach: Provide information about relevant factors or domain-specific rules that influenced the model's decision.

10. **Research and Education:**

    Explanation: Promote research and education around explainable AI.

    Approach: Encourage the AI community to develop techniques and frameworks that enhance interpretability. Educate users about the limitations and possibilities of AI systems.

11. **Ethical Considerations:**

    Explanation: Address ethical concerns related to opacity.

    Approach: Implement mechanisms to identify and mitigate biases, ensuring that decisions made by the black-box model are fair and unbiased.

12. **Regulations and Standards:**

    Explanation: Advocate for regulations that require AI systems to be explainable.

Approach: Support the development of standards that define transparency requirements for AI models in various industries.

13. **Research into Model Internals:**

Explanation: Investigate the inner workings of black-box models.

Approach: Research techniques that allow for understanding complex models, such as visualizing activations, attention mechanisms, or saliency maps.

Addressing the "black-box" nature of certain AI models involves a combination of model selection, interpretation techniques, visualization, education, and ethical considerations. As AI technologies advance, efforts to promote transparency and explainability will become increasingly important for building trustworthy and accountable AI systems that are understood and accepted by users, regulators, and stakeholders.

Balancing Model Performance and Explainability

Balancing model performance and explainability is the practice of finding the right trade-off between the accuracy of complex machine learning models and the transparency of simpler, interpretable models. It's essential for ensuring that AI systems deliver accurate results while also allowing humans to understand how and why those results are generated. Striking this balance is particularly crucial in applications where trust, accountability, and ethical considerations are paramount.

1. **Understanding the Trade-off:**

Explanation: As model complexity increases, performance often improves, but interpretability may decrease.

Approach: Recognize that there is a trade-off between achieving the highest accuracy and maintaining the ability to explain and understand the model's decisions.

2. **Domain and Application Context:**

   Explanation: Consider the domain and application when deciding on the level of explainability required.

   Approach: In applications where transparency and accountability are crucial, opt for more interpretable models, even if they have slightly lower performance.

3. **Use-Case Analysis:**

   Explanation: Assess the specific use case and its impact on the balance between performance and explainability.

   Approach: Determine whether explainability is more critical for legal compliance, ethical considerations, or user trust, and adjust model complexity accordingly.

4. **Model Selection:**

   Explanation: Choose models that inherently balance performance and explainability.

   Approach: Opt for models like decision trees, linear regressions, or rule-based systems that offer reasonable accuracy while being more interpretable.

5. **Feature Selection and Engineering:**

   Explanation: Simplify models by focusing on relevant features and reducing noise.

Approach: Feature selection and engineering can help improve model performance while maintaining interpretability by focusing on the most important input variables.

6. **Regularization Techniques:**

Explanation: Use regularization to control model complexity and prevent overfitting.

Approach: Techniques like L1 or L2 regularization can help balance model performance and prevent models from becoming overly complex.

7. **Ensemble Models:**

Explanation: Combine simpler models into an ensemble to enhance overall accuracy.

Approach: Ensemble methods, like random forests, can improve performance while retaining some level of interpretability through combining multiple base models.

8. **Model Pruning:**

Explanation: Trim unnecessary components of complex models to simplify them.

Approach: Pruning removes parts of the model that do not contribute significantly to its performance, resulting in a more streamlined and comprehensible model.

9. **Interpretability Techniques:**

Explanation: Apply model-agnostic interpretability techniques to complex models.

Approach: Techniques like LIME and SHAP can provide explanations for specific predictions, adding transparency to black-box models without compromising performance.

10. **Documentation and Communication:**

Explanation: Communicate the trade-off to stakeholders and users.

Approach: Document the level of complexity, the model's strengths and limitations, and the rationale behind model choices to manage expectations.

11. **Gradual Complexity Increase:**

Explanation: Start with simpler models and incrementally increase complexity.

Approach: If necessary, begin with an interpretable model and only move to more complex architectures if the performance gains are substantial and justifiable.

12. **User-Centric Approach:**

Explanation: Prioritize user needs and understanding in model development.

Approach: Tailor the level of model complexity and explainability to match the user's background, expertise, and preferences.

Balancing between model performance and explainability requires a thoughtful approach that considers the specific application, stakeholders' needs, and ethical considerations. Striking the right balance ensures that AI systems are not only accurate but also transparent, interpretable, and trustworthy, fostering user trust and enabling responsible AI deployment.

Trade-offs between Model Complexity, Performance, and Explainability

In the field of machine learning, building models involves striking a balance between several important factors: model complexity, performance, and explainability. Each of these aspects contributes to the overall effectiveness and usability of a model, but they often exist in a trade-off relationship, where improving one aspect might come at the cost of another.

Model Complexity

Model complexity refers to how intricate the relationships and patterns within a model are. Complex models can capture intricate patterns, but they might also be more prone to overfitting, which occurs when a model fits the training data too closely and performs poorly on new, unseen data.

Performance

Model performance measures how well a model generalizes its predictions to unseen data. High-performing models exhibit low error rates and accurate predictions. Achieving high performance is a fundamental goal in machine learning, especially for applications like image recognition, natural language processing, and medical diagnosis.

Explainability

"Explainability" refers to the ability to understand and interpret how a model arrives at its decisions. Interpretable models provide clear rationales for their predictions, which enhances user trust and facilitates debugging. Explainability is crucial in domains where transparency, accountability, and ethical considerations are paramount.

Trade-offs: Model Complexity vs. Performance

The trade-off between model complexity and performance is a fundamental consideration in machine learning. As model complexity increases, it often leads to improved performance or accuracy on training data. However, this can come at the cost of reduced interpretability and the risk of overfitting, where the model fits noise in the data rather than the underlying patterns. Simpler models are more interpretable but may sacrifice some performance. Striking the right balance is a critical challenge, and it depends on the specific requirements and constraints of each machine learning task.

- Complex models (e.g., deep neural networks) can achieve high performance by capturing intricate patterns in data.

- However, increased complexity may lead to overfitting, causing the model to perform poorly on new data.

- Achieving high performance often requires a balance between complexity and regularization techniques to prevent overfitting.

Model Complexity vs. Explainability

The trade-off between model complexity and explainability is a central challenge in machine learning. It involves finding the right balance between using complex models that can achieve high predictive performance but are harder to interpret and simpler models that are more interpretable but may sacrifice some predictive accuracy. In many real-world applications, particularly those with high stakes or regulatory requirements, achieving a balance between model complexity and explainability is essential. The choice depends on the specific needs and constraints of the task and the importance of understanding how and why a model makes its predictions.

- Complex models tend to be less interpretable due to their intricate structures and a larger number of parameters.

- As model complexity increases, understanding the decision-making process becomes challenging.

- Simplified models (e.g., linear regression) are more interpretable but might sacrifice some performance.

Performance vs. Explainability

Balancing model performance and explainability is about finding the right trade-off between high accuracy and model transparency. Complex models can offer better performance but are less interpretable, while simpler models are easier to understand but may sacrifice some accuracy. The choice depends on the specific application and the need for insights into the model's decision-making process.

- High-performing models might rely on complex, "black-box" structures that are hard to explain.

- Model performance can sometimes come at the cost of reduced interpretability.

- Achieving a balance involves using techniques like model-agnostic explanations (LIME, SHAP) or hybrid models that combine the strengths of both.

The trade-offs between model complexity, performance, and explainability are central to the design and deployment of machine learning models. Striking the right balance depends on the application's requirements, user expectations, and ethical considerations. Understanding these trade-offs empowers practitioners to make informed decisions about model design and optimization, leading to responsible and effective AI systems.

Conclusion

This chapter covered various aspects of transparency and explainability in AI, including their importance, challenges, methods, tools, real-world examples, and more. We've delved into the significance of these concepts for building user trust, ensuring fairness, and addressing ethical concerns in AI systems. We've explored strategies for achieving transparency and explainability, such as model interpretability techniques, feature importance analysis, and visualization methods.

In the realm of model complexity, we've examined how deep learning and complex AI architectures pose challenges related to understanding, accountability, biases, fairness, and more. The trade-offs between model performance and explainability were explored in depth, emphasizing the ethical considerations that should guide such decisions.

Furthermore, we've discussed the need for explainable AI to prevent biases, discrimination, and social inequalities. We've recognized the value of transparency in enhancing user trust and confidence, while also acknowledging the challenges inherent in deploying complex AI models.

In conclusion, the journey through transparency and explainability in AI reveals their critical role in building responsible, ethical, and trustworthy AI systems. As AI technology continues to evolve, the principles we've discussed serve as guiding lights to ensure that innovation aligns with societal values, user needs, and ethical considerations. The pursuit of transparent and explainable AI fosters a more inclusive, accountable, and positive AI future for all.

CHAPTER 4

Privacy and Security

In the age of digital transformation, where data is often referred to as the "new oil," the importance of safeguarding this invaluable resource cannot be overstated. As artificial intelligence (AI) systems become increasingly integrated into various sectors, from health care to finance, ensuring their security and preserving the privacy of the data they handle becomes paramount. This chapter delves into the intricate world of AI privacy and security, shedding light on the potential vulnerabilities AI systems might have and the tools and techniques available to address them.

The rapid proliferation of AI has brought with it a host of challenges. While AI models can process vast amounts of data to make accurate predictions or classifications, they can also inadvertently expose sensitive information or become targets for malicious attacks. This chapter aims to equip readers with an understanding of these challenges and the means to combat them.

We'll explore the nuances of data and model privacy, learning the potential pitfalls and the techniques used to ensure data remains confidential. The chapter will also delve into the various security measures essential for robust AI systems, highlighting potential threats and the means to mitigate them.

© Avinash Manure, Shaleen Bengani, Saravanan S 2023
A. Manure et al., *Introduction to Responsible AI*,
https://doi.org/10.1007/978-1-4842-9982-1_4

Privacy Concerns in AI

In the modern digital era, AI stands as a transformative force, reshaping industries, enhancing user experiences, and offering unprecedented capabilities. From personalized product recommendations to advanced medical diagnostics, AI's influence is pervasive. However, this immense power is tethered to a fundamental resource: data. As AI systems often operate on vast datasets, encompassing everything from user preferences to personal identifiers, the intersection of AI and privacy has become a focal point of discussion.

Privacy, at its core, pertains to individuals' rights to their personal data. It's about controlling who has access to data, under what conditions, and for what purposes. In the realm of AI, privacy takes on heightened significance. AI models, especially deep learning ones, are data-hungry. They thrive on large datasets, extracting patterns and making predictions. But where does this data come from? Often, it's derived from users—their behaviors, preferences, interactions, and more. This data, while invaluable for AI, is deeply personal to the individuals it represents.

The challenge arises when the insatiable appetite of AI for data meets the intrinsic human right to privacy. AI systems, in their quest for accuracy, can sometimes overstep, delving into data realms that might be considered invasive or overly personal. For instance, an AI system recommending movies based on viewing history is convenient. But what about an AI predicting personal health issues based on search history? The line between helpful and intrusive can be thin, and it's a line that AI developers and users must navigate with care.

Moreover, the decentralized nature of the internet means data can traverse borders, making it accessible from anywhere in the world. This global accessibility further complicates the privacy paradigm. Different cultures and jurisdictions have varied perspectives on privacy, leading to diverse regulations and expectations.

In essence, as AI continues its march forward, intertwining more deeply with our daily lives, the dialogue around privacy becomes not just relevant, but also essential. Balancing the promise of AI with the sanctity of personal privacy is a challenge that developers, policymakers, and users must collectively address.

Potential Threats to Privacy

As AI systems become increasingly sophisticated, the potential threats to privacy have also evolved, becoming more complex and harder to mitigate. While the power to analyze and interpret data is AI's greatest strength, it can also be its most significant weakness when it comes to safeguarding privacy. The following are some of the most pressing threats to privacy in the realm of AI.

Data Breaches and Unauthorized Access

Data breaches pose a significant threat to the privacy of individuals whose information is stored and processed by AI systems. Especially in cloud-based deployments, vulnerabilities can be exploited to gain unauthorized access to sensitive data. For instance, the 2019 security breach at Capital One exposed the personal information of over 100 million customers. The breach was attributed to a misconfigured firewall in a cloud environment, highlighting the potential risks associated with cloud-based AI systems. Such incidents underscore the need for robust security measures, including encryption and multi-factor authentication, to protect the data that AI models use.

Misuse of Personal Data by AI Models

Another growing concern is the misuse of personal data by AI algorithms. Users often provide explicit consent for their data to be used for specific purposes, but AI models can sometimes use this data in ways that were not initially agreed upon. Facial recognition technology serves as a prime example. While users may consent to the use of their images for authentication or tagging in photos, these images can be misused for surveillance or data mining without explicit user consent. This raises ethical concerns and calls for stricter regulations on how AI models can use personal data.

Inadvertent Sharing of Sensitive Information

AI models, particularly those specializing in natural language processing (NLP), can sometimes inadvertently share sensitive information. For example, in health care, AI algorithms are increasingly being used to analyze medical records, predict diseases, and even assist in surgeries. While these applications have numerous benefits, there is also a risk of inadvertent disclosure of patient data. An NLP model trained on a large dataset of medical records could potentially generate outputs that reveal sensitive patient information, such as medical conditions or treatment histories. This not only violates privacy but could also have legal repercussions under laws like the Health Insurance Portability and Accountability Act (HIPAA) in the United States.

Understanding these potential threats is essential for anyone involved in the development, deployment, or utilization of AI systems. Each of these challenges presents unique obstacles to preserving privacy and necessitates specialized techniques and considerations for effective mitigation.

Privacy Attacks in AI Models

In addition to the general privacy concerns outlined earlier, AI models are susceptible to specific types of attacks that pose direct threats to user privacy. These attacks exploit the very mechanisms that make AI powerful, turning them into vulnerabilities. The following are some of the most critical privacy attacks that AI models can fall victim to.

Data Re-identification

While big data anonymization techniques aim to protect individual privacy, AI algorithms have become increasingly proficient at re-identifying this anonymized data. For example, machine learning models can cross-reference anonymized health-care records with publicly available data to re-identify individuals. This is especially concerning when the data is of a sensitive nature, such as medical or financial records.

Inference Attacks

AI models can inadvertently reveal sensitive information through inference attacks. A recommendation system, for example, could unintentionally expose details about a user's personal preferences or even health conditions based on the data it has been trained on and the recommendations it makes. The line between convenience and intrusion is often blurred, making it a challenging issue to address.

Membership Inference Attacks

When an attacker can determine whether a specific data point was part of a machine learning model's training set, it's known as a membership inference attack. This is particularly concerning when the model is trained on sensitive data, such as medical or financial records. The mere knowledge that an individual's data was used in such a model could reveal that the individual has a specific condition or financial status.

Model Inversion Attacks

Model inversion attacks occur when an attacker uses the output of a machine learning model to infer details about the data on which it was trained. For example, if a model predicts the likelihood of a particular disease, an attacker could use the model's output to infer sensitive health information about individuals in the training set.

Understanding the specific types of privacy attacks that AI models are vulnerable to is a critical aspect of responsible AI development and deployment. These attacks not only exploit the technical intricacies of machine learning algorithms but also pose ethical dilemmas that challenge the balance between utility and privacy. As AI continues to advance and integrate more deeply into various sectors, the potential for these privacy attacks will likely increase. Therefore, it is imperative for developers, policymakers, and end users to be aware of these vulnerabilities and to actively seek solutions that mitigate the risks while preserving the transformative benefits of AI technology.

Mitigating Privacy Risks in AI

Navigating the complex landscape of AI privacy risks requires a multi-pronged approach. This section is designed to be a practical guide, offering actionable strategies to mitigate these risks. From technical solutions like data anonymization and differential privacy to ethical considerations such as user consent, we will explore a range of tactics to bolster the privacy of AI models. Each subsection will include code examples for hands-on implementation, allowing you to apply these strategies in real-world scenarios.

Data Anonymization and Encryption

Data anonymization and encryption serve as foundational techniques in the realm of AI privacy. While they are distinct in their operations, both aim to protect sensitive information from unauthorized access and misuse. Data anonymization focuses on transforming the data in a way such that it can no longer be associated with a specific individual. This is particularly useful in mitigating risks such as data breaches and the misuse of personal data by AI models. Techniques like k-anonymity can be employed, where certain attributes are generalized to ensure that each record is indistinguishable from at least $k-1$ other records in the dataset. There are various other methods to anonymize data, such as ℓ-diversity, t-closeness, data masking, and data perturbation. Exploring these is left as an exercise for the reader.

For example, consider a dataset containing age and zip code. Using k-anonymity, the ages could be generalized into age ranges, and zip codes truncated, so that each row would be similar to at least $k-1$ other rows. While there are several ways of k-anonymizing data, here's an example using the ***pandas*** library to achieve this.

```
[In]:
import pandas as pd

# Sample data
df = pd.DataFrame({'Age': [25, 30, 35, 40], 'Zip': [12345,
67890, 12345, 67890]})

# Apply k-anonymity
df['Age'] = pd.cut(df['Age'], bins=[20, 29, 39, 49],
labels=['20-29', '30-39', '40-49'])
df['Zip'] = df['Zip'].apply(lambda x: str(x)[:3])
print(df)

[Out]:
```

	Age	Zip
0	20–29	123
1	30–39	678
2	30–39	123
3	40–49	678

Meanwhile, encryption adds a layer of security by converting data into a code to prevent unauthorized access. It's especially useful when you need to store or transmit the data securely. Advanced Encryption Standard (AES) is one of the most secure encryption algorithms and is widely used for this purpose.

In a real-world scenario, you might encrypt sensitive data before storing it in a database. When you need to use this data for analysis, you decrypt it back into its original form. Here's a Python example using the *cryptography* library to demonstrate AES encryption and decryption.

```
[In]:
from cryptography.fernet import Fernet
# Generate a key
key = Fernet.generate_key()
cipher_suite = Fernet(key)
# Sample data to encrypt
text = "Sensitive Information"
# Encrypt the data
encrypted_data = cipher_suite.encrypt(text.encode())
# Decrypt the data
decrypted_data = cipher_suite.decrypt(encrypted_data).decode()
print(f"Encrypted: {encrypted_data}")
print(f"Decrypted: {decrypted_data}")
```

```
[Out]:
Encrypted: b'gAAAAABk7wx4JXbPBtH_Q5nH52TM67tOO2uvD3g47
pJlayjBObFrBzcDRpgOACEmtrbZuFJ3yMwadoLVya5GPK4j_VpjQRpIzA
CtlyeSrS8bcgEO2NPlUYY='
Decrypted: Sensitive Information
```

In this example, we generate a key using Fernet, a symmetric encryption algorithm included in the cryptography library. The data is then encrypted and stored as `encrypted_data`. When needed, the same key can be used to decrypt the data back to its original form.

By combining data anonymization and encryption, you can create a robust strategy for protecting sensitive information in AI systems, thereby addressing multiple facets of privacy concerns.

Differential Privacy

Differential privacy is a robust privacy-preserving technique that aims to provide means to maximize the accuracy of queries from statistical databases while minimizing the chances of identifying individual entries. It's particularly useful in machine learning where the model's parameters could reveal information about the training data. Differential privacy works by adding a small amount of noise to the data or to the function's output, making it difficult to reverse-engineer the original data from the output.

The concept of differential privacy is often used to mitigate the risks of inference attacks and data re-identification. By adding noise to the data or the query results, it ensures that the removal or addition of a single database item doesn't significantly affect the outcome, thereby preserving individual privacy.

In Python, the **diffprivlib** library from IBM offers a range of differential privacy tools that are compatible with scikit-learn models. Here's a simple example using differential privacy to train a logistic regression model on the Iris dataset:

```
[In]:
from sklearn.datasets import load_iris
from sklearn.linear_model import LogisticRegression
from diffprivlib.models import LogisticRegression as
DPLogisticRegression

# Load dataset
iris = load_iris()
X, y = iris.data, iris.target
```

```
# Train a regular scikit-learn logistic regression model
lr = LogisticRegression()
lr.fit(X, y)

# Train a differentially private logistic regression model
dp_lr = DPLogisticRegression(epsilon=0.1)  # epsilon controls
the privacy budget
dp_lr.fit(X, y)
```

Both models can now be used for prediction, but dp_lr provides differential privacy guarantees.

The parameter epsilon controls the privacy budget: a smaller value provides stronger privacy but may reduce the utility of the data. It's a crucial parameter that needs to be set carefully based on the specific privacy requirements of your application.

Differential privacy is a powerful tool for preserving privacy in AI models, but it's essential to understand its limitations. While it can protect against many types of privacy attacks, it's not a silver bullet and should be part of a broader privacy-preserving strategy.

Secure Multi-Party Computation

Secure multi-party computation (SMPC) is a cryptographic technique that allows multiple parties to collaboratively compute a function over their inputs while keeping the inputs private. In the context of AI and machine learning, SMPC can be used to train a model on a combined dataset that is distributed across multiple locations, without moving the data. This is particularly useful in scenarios where data privacy regulations or business constraints prevent data from being centralized.

SMPC is effective in mitigating risks related to data breaches and unauthorized access. Since the raw data never leaves its original location and only encrypted information is exchanged, the risk of exposing sensitive information is significantly reduced. It also addresses the misuse of personal data by AI models, as the model never has direct access to individual data points—only to encrypted aggregates.

One of the popular Python libraries for implementing SMPC is **PySyft**. It extends PyTorch and TensorFlow to enable multi-party computations. Here's a simplified example using PySyft to perform secure computation:

```
[In]:
import syft as sy
import torch
# Initialize hook to extend PyTorch
hook = sy.TorchHook(torch)
# Create virtual workers
alice = sy.VirtualWorker(hook, id="alice")
bob = sy.VirtualWorker(hook, id="bob")
# Create tensors and send to workers
x = torch.tensor([1, 2, 3, 4, 5])
x_ptr = x.send(alice, bob)
# Perform secure computation
result_ptr = x_ptr + x_ptr
# Get the result back
result = result_ptr.get()
```

In this example, the tensor x is distributed between two virtual workers, Alice and Bob. The computation (x_ptr + x_ptr) is performed securely among the parties, and the result can be retrieved without exposing the individual data.

SMPC is a powerful tool for preserving privacy in AI applications, especially in collaborative environments where data sharing is a concern. However, it's computationally intensive and may not be suitable for all types of machine learning models or large datasets. Therefore, it's crucial to weigh the privacy benefits against the computational costs when considering SMPC for your application.

User Consent and Transparency

User consent is a foundational element in data privacy, especially in AI systems that often require large datasets for training and inference. Obtaining explicit consent is not merely a formality, but rather is a legal and ethical obligation. Consent forms should be designed to be easily understandable, avoiding technical jargon that could confuse users. They should clearly outline the types of data being collected, the specific purposes for which the data will be used, and the duration for which it will be stored.

Moreover, the consent process should be dynamic, allowing users to withdraw their consent at any time. This is particularly important in AI applications that continually evolve and may require additional types of data or may use the data in new ways as the system evolves. In such cases, users should be notified of these changes and be given the option to re-consent.

Transparency goes hand in hand with user consent. While consent gives users a choice, transparency empowers them to make an informed decision. In the context of AI, transparency extends beyond just data collection and usage. It also involves explaining the decision-making process of AI models, especially when these decisions have significant impacts on individuals, such as in health-care diagnostics, financial loan approvals, or job recruitments.

Transparency in AI can be achieved through various means. One approach is to use explainable AI models that allow for the interpretation of model decisions. However, explainability often comes at the cost of model complexity and performance, leading to a trade-off that needs to be carefully managed.

Another avenue for enhancing transparency is through detailed documentation that outlines the AI system's architecture, the data it uses, and the algorithms that power it. This documentation should be easily accessible and understandable to non-experts.

In addition, user-friendly dashboards can be developed to allow users to see in real-time how their data is being used and manipulated by the AI system. These dashboards can also provide users with the ability to control or limit the usage of their data, thereby giving them a tangible sense of control and assurance.

Summary

In this section, we explored a range of strategies for mitigating privacy risks in AI, from data anonymization and encryption to more advanced techniques like differential privacy and secure multi-party computation. We also emphasized the ethical imperatives of user consent and transparency, which serve as foundational elements in any privacy-preserving AI system. These methods, often used in combination, aim to balance the AI's need for data with the individual's right to privacy, thereby fostering responsible AI development and deployment.

Security Concerns in AI

While the previous sections have extensively covered the privacy implications of AI, it's crucial to recognize that security is the other side of the same coin. As AI systems continue to revolutionize everything from health-care diagnostics to financial risk assessments, they also become attractive targets for malicious actors aiming to exploit vulnerabilities for gain or sabotage. The stakes are high; a compromised AI system can not only leak sensitive data but also make erroneous decisions that could have far-reaching consequences, from misdiagnosing medical conditions to causing financial ruin.

The security of AI systems is not just about protecting the data they use but also about ensuring the integrity of the AI models themselves. Sophisticated attacks can manipulate the behavior of AI systems in unpredictable ways, turning them into unwitting accomplices in malicious activities. For example, an adversarial attack could subtly alter the data fed into an AI-based security system, causing it to overlook unauthorized intrusions.

Moreover, the security landscape for AI is complicated by the very nature of machine learning, which often involves decentralized data sources and multiple stakeholders, from data scientists to end users. This multifaceted environment creates numerous points of vulnerability that need to be safeguarded rigorously.

In this section, we will dissect the various types of security threats to which AI systems are susceptible and explore comprehensive strategies to mitigate these risks. By the end of this section, you'll have a well-rounded understanding of how to build AI systems that are not just smart, but also secure and trustworthy.

Potential Threats to Security

Security threats to AI systems can range from data tampering to model manipulation, and they pose significant risks not only to the AI system itself but also to the people and organizations that rely on it. In this section, we will explore some of the most pressing security threats that AI systems face today.

Adversarial Attacks

One of the most talked-about security threats in AI is the adversarial attack, where small, carefully crafted changes to the input data can lead to incorrect outputs. For instance, adding a specific kind of noise to an image could cause a facial recognition system to misidentify a person. These attacks are particularly concerning because they can be hard to detect and can compromise the integrity of the AI system's decision making.

Data Poisoning

Data poisoning involves manipulating the training data so that the AI model learns incorrect behaviors. This is a more insidious form of attack because it happens during the model-training phase, making it difficult to identify later. For example, inserting fake news articles into a dataset used to train a news classification algorithm could lead the model to make biased or incorrect classifications.

Model Inversion and Extraction

In a model inversion attack, an attacker uses the outputs of a machine learning model to reconstruct sensitive information about the training data. Model extraction attacks aim to replicate a proprietary model by querying it multiple times. Both of these attacks can lead to intellectual property theft and compromise user privacy.

Evasion Attacks

These attacks occur when an adversary manipulates input data during the inference stage to receive a specific output, effectively "evading" the intended function of the model; for example, altering the features of a malware file to make it appear benign to an AI-based antivirus system.

Backdoor Attacks

In a backdoor attack, the attacker introduces a hidden pattern or "backdoor" into the training data, which the model then learns. When the model encounters this pattern in future data, it produces an output specified by the attacker, essentially giving them control over the model's decisions for those instances.

Understanding these potential security threats is the first step in creating robust AI systems. Each of these threats presents unique challenges that require specialized techniques for mitigation, which we will explore in the next section.

Mitigating Security Risks in AI

Securing AI systems is a multifaceted endeavor that demands a comprehensive toolkit of strategies. This section serves as a hands-on guide, detailing actionable measures to fortify the security of your AI models against various threats. We'll delve into a spectrum of techniques, from defensive mechanisms against adversarial attacks to robust data-validation methods and real-time monitoring solutions. Each subsection will not only explain the theoretical underpinnings but also provide code examples for practical implementation, equipping you with the skills needed to safeguard AI systems in real-world applications.

Defense Mechanisms against Adversarial Attacks

One of the most effective ways to defend against adversarial attacks is through **adversarial training**. This involves augmenting the training dataset with adversarial examples and retraining the model. By doing so, the model becomes more robust and less susceptible to adversarial manipulation.

Here's a simple example that uses TensorFlow to demonstrate adversarial training:

```
[In]:
import tensorflow as tf

# Create a simple model
model = tf.keras.Sequential([
    tf.keras.layers.Flatten(input_shape=(28, 28)),
```

```
    tf.keras.layers.Dense(128, activation='relu'),
    tf.keras.layers.Dense(10)
])

# Compile the model
model.compile(optimizer='adam',
              loss=tf.keras.losses.SparseCategoricalCrossentrop
y(from_logits=True),
              metrics=['accuracy'])

# Generate adversarial examples by adding random noise
train_images_adv = train_images + 0.01 *
tf.random.normal(shape=train_images.shape)

# Combine original and adversarial examples for training
train_images_combined = tf.concat([train_images, train_images_
adv], axis=0)
train_labels_combined = tf.concat([train_labels, train_
labels], axis=0)

# Train the model
model.fit(train_images_combined, train_labels_combined,
epochs=5)
```

Gradient masking is another technique used to defend against adversarial attacks. It involves modifying the model's architecture or training process to make it difficult for attackers to compute the gradients needed to generate adversarial examples. This can be achieved through techniques like dropout, L1/L2 regularization, or even custom loss functions that penalize certain types of adversarial perturbations. The following example showcases how to build a model with L2 regularization:

[In]:
```
from tensorflow.keras import regularizers
```

```
# Create a simple model with L2 regularization
model = tf.keras.Sequential([
    tf.keras.layers.Flatten(input_shape=(28, 28)),
    tf.keras.layers.Dense(128, activation='relu', kernel_
regularizer=regularizers.l2(0.01)),
    tf.keras.layers.Dense(10)
])

# Compile the model
model.compile(optimizer='adam',
              loss=tf.keras.losses.SparseCategoricalCrossentropy
              (from_logits=True),
              metrics=['accuracy'])

# Train the model
model.fit(train_images, train_labels, epochs=5)
```

The regularization term helps in constraining the model's capacity, making it less likely to fit to the noise introduced by adversarial perturbations.

Feature squeezing is a pre-processing step that reduces the search space available to an adversary. By reducing the depth, color, or even spatial resolution of the input data, the model becomes less sensitive to the small perturbations in adversarial attacks. This can be particularly useful in image classification tasks.

Here's a simple example of feature squeezing by reducing the color depth of images:

```
[In]:
import numpy as np

# Function to reduce color depth
def reduce_color_depth(image, bit=4):
```

```
    return np.floor(image * (2**bit - 1)) / (2**bit - 1)
# Apply feature squeezing to training images
train_images_squeezed = reduce_color_depth(train_images)
```

By incorporating these additional defensive mechanisms, you can further bolster your AI system's resilience against adversarial attacks. These methods can be used in isolation or combined for a more robust defense strategy. The key is to understand the specific vulnerabilities of your model and the types of attacks it is most likely to face, and then tailor your defensive approach accordingly.

Model Hardening

Model hardening is a set of techniques aimed at making AI models more resilient to various types of attacks, including model extraction and reverse engineering. These techniques often involve adding layers of complexity or security checks that make it difficult for attackers to manipulate the model or gain unauthorized access. For instance, rate limiting can be implemented to restrict the number of queries to the model, making it harder for attackers to reverse engineer it. Another approach is to use obfuscation techniques that make the model's internal workings less transparent, thereby complicating any attempts to tamper with it. Obfuscation can be achieved through various means, such as adding noise to the data, ensembling multiple models, or even altering the architecture in subtle ways that don't affect performance but add complexity. For example, you could use ensemble methods that combine predictions from multiple base models, each with its own layer of added noise or data transformations. Here's an example using **scikit-learn** that combines Gaussian noise and ensembling:

```
[In]:
from sklearn.base import BaseEstimator, TransformerMixin
from sklearn.pipeline import Pipeline
```

```python
from sklearn.ensemble import RandomForestClassifier,
VotingClassifier
from sklearn.linear_model import LogisticRegression
from sklearn.svm import SVC
import numpy as np

# Custom transformer to add Gaussian noise
class GaussianNoise(BaseEstimator, TransformerMixin):
    def __init__(self, noise_level=0.01):
        self.noise_level = noise_level

    def fit(self, X, y=None):
        return self

    def transform(self, X):
        noise = np.random.normal(0, self.noise_level, X.shape)
        return X + noise

# Sample data
X_train = np.random.rand(100, 10)
y_train = np.random.randint(2, size=100)

# Create pipelines with obfuscation
pipeline1 = Pipeline([
    ('noise_adder', GaussianNoise(0.01)),
    ('classifier', LogisticRegression())
])

pipeline2 = Pipeline([
    ('noise_adder', GaussianNoise(0.02)),
    ('classifier', RandomForestClassifier())
])

pipeline3 = Pipeline([
    ('noise_adder', GaussianNoise(0.03)),
```

```
    ('classifier', SVC())
])

# Create an ensemble of the pipelines
ensemble_clf = VotingClassifier(estimators=[
    ('lr', pipeline1),
    ('rf', pipeline2),
    ('svc', pipeline3)
], voting='hard')

# Train the ensemble model
ensemble_clf.fit(X_train, y_train)

# Sample test data
X_test = np.random.rand(5, 10)

# Make predictions
y_pred = ensemble_clf.predict(X_test)
print("Predictions:", y_pred)
```

Input Filtering for Evasion Attacks

Input filtering for evasion attacks is a crucial defensive measure that aims
to identify and neutralize malicious inputs designed to deceive AI models.
Evasion attacks often involve slightly altering the input data in a way that
is imperceptible to humans but can cause the model to make incorrect
predictions or classifications. By implementing input filtering, you can
scrutinize incoming data for any anomalies or suspicious patterns that
could indicate an evasion attack.

For example, in the context of image recognition models, input filtering
could involve checking for pixel values that deviate significantly from the
expected range or patterns. Similarly, for text-based models, you could
look for unusual character sequences or unexpected language constructs
that are not typical of genuine inputs.

Here's a simple Python code snippet that demonstrates basic input filtering for a text-based model using scikit-learn:

```
[In]:
from sklearn.feature_extraction.text import CountVectorizer
from sklearn.naive_bayes import MultinomialNB
import re

# Sample data and labels
X_train = ["hello", "world", "hello world"]
y_train = [1, 0, 1]

# Train a simple model
vectorizer = CountVectorizer()
X_train_vectorized = vectorizer.fit_transform(X_train)
clf = MultinomialNB()
clf.fit(X_train_vectorized, y_train)

# Input filtering function
def filter_input(input_text):
    # Remove any unusual characters
    filtered_text = re.sub(r'[^a-zA-Z0-9\s]', '', input_text)
    return filtered_text

# Test the model with and without input filtering
test_input = "hello w@rld!!"
filtered_test_input = filter_input(test_input)

# Transform the test input
test_vectorized = vectorizer.transform([filtered_test_input])

# Make a prediction
prediction = clf.predict(test_vectorized)
```

Backdoor Detection and Removal

Backdoor detection and removal is an essential aspect of securing AI models, especially those that are trained on data from multiple sources or are part of a larger, interconnected system. Backdoors are malicious functionalities that are secretly inserted into the model during the training phase. These backdoors can be activated later to manipulate the model's behavior, often in ways that are detrimental to its intended function.

For example, a backdoor could be inserted into a facial recognition system that allows unauthorized users to gain access simply by wearing a specific pattern on their clothing. Detecting such backdoors involves rigorous testing and validation of the model's behavior under various conditions, as well as scrutinizing the training data for anomalies or suspicious patterns.

Here's a simple Python example using scikit-learn to demonstrate a basic method for backdoor detection:

```
[In]:
from sklearn.datasets import make_classification
from sklearn.ensemble import RandomForestClassifier
import numpy as np

# Create a synthetic dataset
X, y = make_classification(n_samples=1000, n_features=20,
random_state=42)

# Insert a backdoor: if the sum of features is greater than a
threshold, label is set to 1
threshold = 10
for i in range(len(X)):
    if np.sum(X[i]) > threshold:
        y[i] = 1

# Train a model
```

```
clf = RandomForestClassifier()
clf.fit(X, y)

# Backdoor detection: Check feature importance
feature_importances = clf.feature_importances_
suspicious_features = np.where(feature_importances >
np.mean(feature_importances) + 2 * np.std(feature_importances))

# If suspicious features are found, further investigation
is needed
if len(suspicious_features[0]) > 0:
    print(f"Suspicious features detected: {suspicious_
features}")
```

In this example, we use feature importance as a simple metric to detect suspicious behavior in the model. If certain features have an unusually high importance, it could be an indicator of a backdoor. Further investigation would then be required to confirm the presence of a backdoor and to remove it.

Backdoor detection and removal is a complex task that often requires specialized tools and expertise. However, being aware of the possibility and taking steps to detect and remove backdoors can significantly enhance the security of your AI systems.

Monitoring and Auditing

Monitoring and auditing are essential practices for maintaining the privacy and security of AI systems. These processes involve continuously tracking the behavior of AI models and the data they process, as well as periodically reviewing these activities to ensure they adhere to privacy guidelines and regulations.

Monitoring involves real-time tracking of data access, model predictions, and other activities. It can help in quickly identifying any unusual behavior that could indicate a privacy breach or vulnerability. For instance, sudden spikes in data access or unexpected model predictions could be flagged for further investigation.

Auditing, on the other hand, is a more comprehensive review that usually takes place at regular intervals. It involves a detailed examination of logs, configurations, and other records to ensure that the system is compliant with privacy policies and regulations. Auditing can also include penetration testing, where ethical hackers attempt to exploit vulnerabilities in the system to assess its robustness.

In a real-world scenario, monitoring usually involves very sophisticated tools and practices. For instance, you might use a monitoring solution like Prometheus to collect metrics from your AI models and Grafana to visualize those metrics. Auditing could involve using specialized software that tracks data lineage, access logs, and model decisions over time.

Monitoring and auditing are not just technical requirements but are also often legally mandated, especially in sectors like health care and finance. They provide both a first line of defense against privacy breaches and a means of accountability, ensuring that any lapses in privacy can be traced, understood, and rectified.

Summary

Securing AI models is a multifaceted endeavor that goes beyond just implementing robust algorithms. From defending against adversarial attacks through techniques like gradient masking and adversarial training, to detecting and removing backdoors from models, each strategy plays a crucial role in fortifying your AI systems. Model-hardening techniques like obfuscation add layers of complexity that deter reverse engineering, while monitoring and auditing provide the necessary oversight to detect and respond to anomalies swiftly. By adopting a comprehensive approach that integrates these various strategies, you can significantly enhance the security posture of your AI models, making them resilient against a wide array of potential threats.

Conclusion

In this chapter, we navigated the complex landscape of privacy and security challenges that come with the increasing integration of artificial intelligence in various sectors. We began by dissecting the potential threats to privacy, such as data breaches, unauthorized access, and misuse of personal data. We then transitioned into actionable strategies for mitigating these risks, covering a wide array of techniques from data anonymization and encryption to differential privacy and user consent.

On the security side, we scrutinized the types of vulnerabilities AI models are exposed to, including adversarial attacks, data poisoning, and backdoors. We offered a detailed guide on how to fortify these systems against such threats, discussing methods like model hardening, data validation, and real-time monitoring and alerting. Now at the end of this chapter, you should be well equipped with a holistic understanding of the challenges and solutions to ensuring that AI models are both powerful and ethical, secure yet transparent.

CHAPTER 5

Robustness and Reliability

While much of the excitement around artificial intelligence (AI) focuses on achieving high accuracy and impressive performance, the importance of robustness and reliability often gets overlooked. These foundational elements serve as the bedrock upon which truly effective and safe AI systems are built. In a world increasingly dependent on automated decision making, the robustness of an AI model—its ability to perform consistently under varying conditions—and its reliability—its capacity to produce trustworthy and repeatable results—are not just desirable traits; they are absolute necessities.

The real-world applications of AI are far more complex and unpredictable than controlled lab environments. AI models are deployed in health care for diagnostics, in autonomous vehicles for navigation, in finance for fraud detection, and in many other critical sectors. In these settings, a model's failure to be robust and reliable can have consequences that range from financial losses to endangering human lives. Therefore, understanding how to build and validate robust and reliable AI models is not just an academic exercise but a practical necessity.

This chapter aims to delve deep into the intricacies of ensuring robustness and reliability in AI systems. We will explore the challenges and pitfalls that developers and data scientists face when transitioning from a prototype to a production-level model. The chapter will cover a range of

A. Manure et al., *Introduction to Responsible AI*,
https://doi.org/10.1007/978-1-4842-9982-1_5

techniques, from data preprocessing methods that enhance robustness to advanced algorithms that ensure reliability, complete with practical code examples to demonstrate these techniques, offering a hands-on approach to understanding and implementing them.

Concepts of Robustness and Reliability

Robustness in AI refers to a system's resilience and ability to maintain performance when faced with adverse conditions or unexpected inputs. This quality is paramount, especially as AI models are deployed in dynamic environments where they encounter varied data patterns, noise, or even deliberate attempts to deceive or manipulate them, such as adversarial attacks. For instance, a robust image-recognition system should be able to accurately identify objects in an image even if the image is partially obscured, distorted, or presented in varying lighting conditions. The essence of robustness lies in ensuring that AI systems can handle uncertainties, outliers, and anomalies without significant degradation in their performance.

Reliability, on the other hand, emphasizes the consistent performance of an AI system over time and across different scenarios. It's about trustworthiness and the assurance that the system will function as expected every time it's called upon. In sectors like health care or finance, where decisions made by AI can have profound implications, reliability becomes paramount. For example, a medical diagnostic AI tool must consistently and accurately identify diseases from patient data over extended periods, irrespective of the volume or the source of the data. Reliability ensures that users can depend on AI systems to deliver predictable and stable outcomes, reinforcing trust and facilitating broader adoption of AI technologies.

Importance in AI Systems

In today's digital age, AI systems are deeply embedded in myriad applications, from health-care diagnostics to financial forecasting, and from autonomous vehicles to content recommendation. The vast influence of AI across sectors underscores the paramount importance of robustness and reliability in these systems. Let's delve into why these attributes are so crucial in AI applications:

Health Care: In the realm of medical imaging, AI models assist radiologists in detecting tumors or anomalies in X-rays and MRI scans. A robust AI system can effectively handle low-quality images or those with unexpected artifacts. However, if the system lacks robustness, it might misinterpret these artifacts, leading to misdiagnoses. Reliability ensures consistent diagnostic suggestions across different scans and patients. An unreliable system might miss a tumor in one patient while detecting it in another with a similar condition, potentially resulting in untreated conditions or unnecessary medical procedures.

Autonomous Vehicles: Self-driving cars rely heavily on AI to interpret their surroundings and make decisions. A non-robust AI might misinterpret a puddle as a pothole, leading to unnecessary evasive maneuvers. Reliability ensures that the car's AI system consistently recognizes and reacts to stop signs, traffic lights, and other signals. An unreliable system might occasionally ignore a stop sign, posing significant safety risks to passengers and pedestrians alike.

Finance: AI-driven algorithms in finance predict stock market movements, assess loan eligibility, and detect fraudulent transactions. Without robustness, the algorithm might misinterpret market anomalies, leading to significant financial losses. Reliability ensures consistent AI decisions when analyzing vast datasets. An unreliable model might approve a high-risk loan, exposing financial institutions to potential defaults.

E-commerce and Content Platforms: Platforms like Netflix or Amazon use AI for recommendations. Without robustness, a temporary change in a user's browsing pattern, like a guest using their account, might skew recommendations for weeks. Reliability ensures users consistently receive high-quality recommendations. An unreliable system might suggest irrelevant products, leading to decreased sales and user dissatisfaction.

Smart Assistants: Devices like Alexa or Siri interpret and respond to user commands. Without robustness, they might misinterpret commands due to background noise or varied accents, leading to incorrect actions. Reliability ensures consistent and accurate device responses. An unreliable assistant might set an alarm for the wrong time or play an undesired song, diminishing user experience.

In all these scenarios, the absence of robustness can lead to AI systems' making incorrect decisions when faced with unexpected data or conditions, often resulting in financial losses, eroded user trust, or even threats to safety. Lack of reliability makes AI performance unpredictable, further eroding trust and potentially leading to repeated errors in decision making. In critical applications like health care or autonomous driving, these shortcomings can have dire consequences, ranging from financial repercussions to threats to human life.

Metrics for Measuring Robustness and Reliability

To ensure that AI systems are both robust and reliable, it's crucial to have quantitative metrics that can assess these attributes. These metrics provide a standardized way to evaluate and compare different models, guiding the development and deployment of AI systems.

Robustness Metrics

Perturbation tolerance evaluates the model's performance under small input perturbations. A model is considered robust if its predictions remain consistent even when the input data is slightly altered.

Given an input x and its perturbed version x', the perturbation tolerance is defined as the difference in model predictions:

$$Tolerance = |f(x) - f(x')|$$

Where f is the model's prediction function.

Adversarial robustness measures the model's resilience to adversarial attacks, where the input is deliberately modified to mislead the model. Given an adversarial input x_{adv} crafted to maximize the prediction error, the adversarial robustness is the difference between predictions on the original and adversarial input:

$$Robustness = |f(x) - f(x_{adv})|$$

Out-of-distribution (OOD) detection evaluates the model's ability to recognize and handle inputs that are not from the training distribution. Area Under the Receiver Operating Characteristic Curve (AUROC) is commonly used for OOD detection. A higher AUROC indicates better OOD detection.

Reliability Metrics

Consistency measures the model's ability to produce consistent predictions over multiple runs, especially in stochastic models.

Given multiple predictions $f_1(x)f_2(x)f_3(x)...f_n(x)$ for the same input x over different runs, the consistency is the variance of these predictions:

$$Consistency = \frac{1}{f_1(x)f_2(x)f_3(x)...f_n(x)}$$

Therefore, a lower variance indicates higher consistency.

Mean time between failures (MTBF) is a metric borrowed from traditional reliability engineering and represents the average time between system failures. In the context of AI systems, a "failure" could be defined as an incorrect prediction, an API error or any other undesirable outcome.

$$MTBF = \frac{total\ operating\ time}{number\ of\ failures}$$

In probabilistic models, **Coverage** represents the proportion of instances for which the model's predicted confidence interval contains the true value. A model with high coverage is considered more reliable.

Incorporating these metrics into the evaluation process ensures a comprehensive assessment of AI models in terms of robustness and reliability. By quantifying these attributes, developers and stakeholders can make informed decisions about model deployment and further refinement.

Challenges in Achieving Robustness

So far, we've understood that it is important for AI models to be robust. However, achieving robustness is not straightforward. As AI models become more complex, they are exposed to various challenges that can undermine their robustness. From adversarial inputs that intentionally mislead models to the pitfalls of overfitting where models perform well on training data but poorly on new data, there are numerous obstacles to consider. This section will explore these challenges in depth, highlighting the vulnerabilities in AI systems and the implications of these challenges in practical scenarios.

Sensitivity to Input Variations

Artificial intelligence models, particularly deep learning architectures, are known for their capacity to capture intricate patterns in data. While this capability is advantageous for tasks like image recognition or natural language processing, it also makes these models highly sensitive to slight variations in their input data. This sensitivity can be a double-edged sword, leading to unexpected and undesirable outcomes in certain scenarios.

The high-dimensional spaces in which deep learning models operate allow them to discern patterns that might be invisible to simpler models or human observers. However, this granularity means that even minute changes in input can traverse vastly different paths through this space, leading to divergent outputs. For instance, in image classification tasks, changing a few pixels might cause a model to misclassify an image, even if the change is imperceptible to the human eye.

One of the most notable manifestations of this sensitivity is in the form of adversarial attacks. In these attacks, malicious actors introduce carefully crafted perturbations to the input data, causing the model to make incorrect predictions or classifications. These perturbations are typically so subtle that they don't alter the human interpretation of the data, but they can drastically affect the model's output. For example, an adversarial attack on an image recognition system might involve adding a slight, almost invisible noise to an image of a cat, causing the model to misclassify it as a dog.

To combat the challenges posed by sensitivity to input variations, researchers and practitioners employ various techniques. Data augmentation, where the training data is artificially expanded by introducing controlled variations, is a common approach. This helps the model become less sensitive to such changes. Additionally, techniques like adversarial training, where the model is trained on both the original and the adversarially perturbed data, can enhance the model's resilience against such attacks.

Let's look at an example to perform data augmentation on the CIFAR-10 image classification dataset:

```
[In]:
import torch
import torchvision
import torchvision.transforms as transforms

# Define data augmentation transforms for the training set
transform_train = transforms.Compose([
    transforms.RandomHorizontalFlip(),
    transforms.RandomCrop(32, padding=4),
    transforms.ToTensor(),
    transforms.Normalize((0.5, 0.5, 0.5), (0.5, 0.5, 0.5)),
])

# Define transforms for the test set
transform_test = transforms.Compose([
    transforms.ToTensor(),
    transforms.Normalize((0.5, 0.5, 0.5), (0.5, 0.5, 0.5)),
])
```

```
# Load CIFAR-10 dataset
trainset = torchvision.datasets.CIFAR10(root='./data',
train=True, download=True, transform=transform_train)
trainloader = torch.utils.data.DataLoader(trainset, batch_
size=32, shuffle=True)

testset = torchvision.datasets.CIFAR10(root='./data',
train=False, download=True, transform=transform_test)
testloader = torch.utils.data.DataLoader(testset, batch_
size=32, shuffle=False)

# Define a simple CNN model (omitted for brevity)

# Training loop (omitted for brevity)
```

In this example, we're using the CIFAR-10 dataset and applying a random horizontal flip and random cropping as data augmentation techniques for the training set. These techniques introduce variability in the training data, helping the model become more robust to slight changes in input.

Model Overfitting

Overfitting is a pervasive challenge in machine learning, where a model becomes so attuned to the training data that it captures not just the underlying patterns but also its noise and outliers. This often results in subpar performance on unseen or new data.

Central to understanding overfitting is the bias–variance trade-off (Figure 5-1). On one side, "bias" refers to the error due to overly simplistic assumptions in the learning algorithm. A high-bias model might miss intricate patterns in the data, leading to systematic errors regardless of the dataset size. On the other side, "variance" refers to the error due to excessive complexity in the learning algorithm. A high-variance model reacts too much to fluctuations in the training data, often capturing noise.

In the context of AI robustness, a model with high variance (low bias) might be easily fooled by slightly altered inputs, making it less dependable in real-world scenarios. Striking the right balance between bias and variance is essential, as it ensures models are both accurate on their training data and generalize well to new, unseen data.

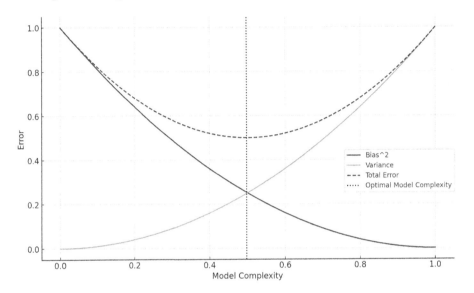

Figure 5-1. *Bias–variance trade-off*

Several factors can contribute to overfitting:

- **Complex Models:** Deep neural networks with a large number of parameters are more prone to overfitting, especially when the amount of training data is limited.

- **Limited Data:** Insufficient training data can lead to a model that's too closely adapted to that data and fails to generalize well.

- **Noisy Data:** If the training data contains errors or noise, a model might learn these as patterns, leading to overfitting.

Here are some ways overfitting can be addressed:

- **Regularization:** For linear models, techniques like Ridge (L2) and Lasso (L1) regularization can be used. These techniques add a penalty to the loss function, discouraging the model from assigning too much importance to any single feature. Having smaller coefficients leads to a simpler, more stable, and generalized model.

- **Pruning:** For decision trees, pruning can be used to remove parts of the tree that do not provide power in predicting target values.

- **Data Augmentation:** By applying various transformations to the original training data, data augmentation artificially increases the data's size and diversity. This enhanced variability helps the model to generalize better across unseen data, thereby reducing the likelihood of overfitting.

- **Feature Selection:** Reducing the number of input features can help in preventing overfitting. Techniques like backward elimination, forward selection, and recursive feature elimination can be used.

- **Increase Training Data:** If feasible, increasing the amount of training data can help in improving the generalization of the model.

- **Cross-Validation:** This involves partitioning the training data into subsets, training the model on some of those subsets, and validating it on others. It helps in ensuring that the model generalizes well.

- **Ensemble Methods:** Techniques like bagging and boosting can be used to combine multiple models to improve generalization.

Here is an example using Ridge regularization with sklearn:

```
[In]:
from sklearn.linear_model import Ridge
from sklearn.datasets import load_boston
from sklearn.model_selection import train_test_split
from sklearn.metrics import mean_squared_error

# Load dataset
data = load_boston()
X_train, X_test, y_train, y_test = train_test_split(data.data,
data.target, test_size=0.2, random_state=42)

# Apply Ridge regularization
ridge = Ridge(alpha=1.0)
ridge.fit(X_train, y_train)

# Predict and evaluate
y_pred = ridge.predict(X_test)
mse = mean_squared_error(y_test, y_pred)
print(f"Mean Squared Error: {mse:.2f}")
```

In this example, we've used the Ridge regression model from sklearn on the Boston housing dataset. The alpha parameter controls the strength of the regularization. The greater the value of alpha, the stronger the regularization. Adjusting this parameter can help in finding the right balance between fitting the training data and generalizing to new data, thereby preventing overfitting.

Outliers and Noise

Data is rarely perfect. Outliers—data points that deviate significantly from other observations—can skew model training. For example, in a dataset of house prices, an abnormally high value due to data entry error can distort the model's understanding. Similarly, noise—or random fluctuations in data—can be detrimental. Consider medical imaging: a model trained on noisy MRI scans might misdiagnose patients, with grave consequences.

Outliers and noise can cause the following issues in AI systems:

- **Skewed Model Training:** Outliers can disproportionately influence model parameters. For instance, in regression models, a single outlier can drastically change the slope of the regression line, leading to inaccurate predictions. This phenomenon is especially pronounced in models that rely on mean and variance, as outliers can skew these metrics.

- **Reduced Model Accuracy:** Noise introduces randomness into the data. When models are trained on noisy data, they might end up capturing this randomness, mistaking it for genuine patterns. This can lead to overfitting, where the model performs well on the training data but poorly on new, unseen data.

- **Compromised Model Interpretability:** Outliers can lead to unexpected model behaviors. For instance, a clustering algorithm might create an entire cluster for a few outliers, leading to misleading interpretations. Similarly, noise can obscure genuine patterns, making it harder to discern what the model has truly learned.

Here are some ways these issues can be addressed:

- **Data Cleaning:** This is the first line of defense. A thorough exploratory data analysis (EDA) can help in visualizing and identifying outliers. Techniques like scatter plots, box plots, and histograms can be instrumental. Once identified, outliers can be removed, capped, or transformed, depending on the context.

- **Robust Scaling:** Traditional scaling methods, like Min-Max scaling or Standard scaling, are sensitive to outliers. Robust scaling techniques, on the other hand, scale features using statistics that are not influenced by outliers, ensuring that the scaled data retains its original distribution.

- **Statistical Methods:** The IQR (Interquartile Range) method, Z-score, and the Modified Z-score are popular techniques to identify and handle outliers. These methods provide a statistical basis for determining which data points deviate significantly from the expected norm.

- **Noise Reduction Techniques:** Smoothing methods, like moving averages, can help reduce noise in time-series data. In image data, filters like Gaussian filters or median filters can be applied to reduce noise.

- **Regularization:** Regularization techniques, like L1 (Lasso) and L2 (Ridge) regularization, add a penalty to the model parameters, ensuring that the model does not fit the noise in the data. This can make models less sensitive to outliers and noise.

Let's demonstrate how to handle outliers using the IQR method and then apply robust scaling on the Boston housing dataset.

```
[In]:
from sklearn.datasets import load_boston
from sklearn.model_selection import train_test_split
from sklearn.preprocessing import RobustScaler

# Load dataset
data = load_boston()
X = data.data
y = data.target

# Split data
X_train, X_test, y_train, y_test = train_test_split(X, y,
test_size=0.2, random_state=42)

# Identify and remove outliers using IQR
Q1 = np.percentile(X_train, 25, axis=0)
Q3 = np.percentile(X_train, 75, axis=0)
IQR = Q3 - Q1
lower_bound = Q1 - 1.5 * IQR
upper_bound = Q3 + 1.5 * IQR

# Only keep rows without outliers
mask = (X_train >= lower_bound) & (X_train <= upper_bound)
X_train = X_train[mask.all(axis=1)]
y_train = y_train[mask.all(axis=1)]

# Apply Robust Scaling
scaler = RobustScaler()
X_train_scaled = scaler.fit_transform(X_train)
X_test_scaled = scaler.transform(X_test)
```

In this example, we first identify and remove outliers using the IQR method. Then, we apply robust scaling to make the model less sensitive to any remaining outliers in the data.

Transferability of Adversarial Examples

Adversarial examples, as previously discussed, are input samples that have been perturbed slightly to deceive machine learning models. These perturbations, often imperceptible to humans, can cause models to make incorrect predictions. One of the most intriguing and concerning properties of adversarial examples is their transferability. This means that an adversarial example crafted to fool one model can often deceive another model, even if the second model has a different architecture or was trained on different data.

The phenomenon of transferability suggests that different models capture similar decision boundaries, especially in the vicinity of the data points. When an adversarial example pushes a data point across the decision boundary in one model, there's a good chance it will have a similar effect in another model. This property is especially concerning in the context of "black-box" attacks, where an attacker doesn't have direct access to a model's parameters or architecture but can still generate adversarial examples using a surrogate model. Once crafted, these adversarial examples can be used to attack the target model.

This has several implications, such as the following:

- **Wider Attack Surface:** Transferability means that even if an attacker doesn't have access to the target model, they can still craft effective adversarial examples using a different model they do have access to. This broadens the potential attack surface significantly.

- **Challenges in Defense:** Defending against adversarial attacks becomes more challenging due to transferability. Even if a model is robust against direct attacks, it might still be vulnerable to adversarial examples crafted using other models.

- **Model Ensemble Vulnerability:** Even ensemble methods, which combine predictions from multiple models to improve accuracy, are not immune. If adversarial examples are transferable across individual models, they can potentially deceive the ensemble as a whole.

However, there are several ways of addressing this challenge. Some of them follow:

- **Adversarial Training:** One of the most effective ways to combat adversarial examples is by incorporating them in the training process. By training models on both original and adversarial data, they become more robust to such perturbations. However, this method can be computationally expensive.

- **Input Preprocessing:** Techniques like image denoising, compression, or smoothing can be applied to input data to reduce the effectiveness of adversarial perturbations.

- **Randomized Defenses:** Introducing randomness into the model, either in its architecture or during inference, can make it harder for adversarial examples to transfer. For instance, randomly dropping out neurons or using stochastic activation functions can increase resilience.

- **Model Diversity:** Encouraging diversity in model architectures and training datasets can reduce the transferability of adversarial examples. If models have distinct decision boundaries, an adversarial example effective against one might not work against another.

Here's a simple example using sklearn and the Fast Gradient Sign Method (FGSM) to generate adversarial examples and use them to make our model more robust against them:

```
[In]:
import numpy as np
from sklearn import datasets
from sklearn.model_selection import train_test_split
from sklearn.ensemble import RandomForestClassifier
from sklearn.metrics import accuracy_score

# Load dataset
data = datasets.load_iris()
X = data.data
y = data.target

# Split the dataset
X_train, X_test, y_train, y_test = train_test_split(X, y,
test_size=0.3, random_state=42)

# Train a RandomForest classifier
clf = RandomForestClassifier().fit(X_train, y_train)

# Define a function to create adversarial examples using FGSM
def fgsm_attack(data, labels, classifier, epsilon):
    # Calculate the gradient of the loss w.r.t the input data
    gradient = classifier.decision_path(data)[1].toarray()
    # Create adversarial examples by adding the sign of the
    gradient multiplied by epsilon
```

```
    adversarial_data = data + epsilon * np.sign(gradient)
    return adversarial_data

# Generate adversarial examples
epsilon = 0.1
X_adversarial_train = fgsm_attack(X_train, y_train, clf,
epsilon)

# Retrain the model with the adversarial examples
clf_defended = RandomForestClassifier().fit(X_adversarial_
train, y_train)

# Test the accuracy of the defended model on the original
test set
accuracy_defended = accuracy_score(y_test, clf_defended.
predict(X_test))

print(f"Accuracy of defended RandomForest on original test set:
{accuracy_defended}")
```

In this example, we first train a RandomForest classifier. We then use the Fast Gradient Sign Method (FGSM) to generate adversarial examples from the training set. We retrain the RandomForest classifier with these adversarial examples. The expectation is that the defended model will perform better against adversarial attacks, including transferable adversarial examples, compared to the original model.

Challenges in Ensuring Reliability

The dynamic nature of real-world data, evolving user behaviors, and changing environments can all influence an AI model's performance. A model that was once deemed state-of-the-art can quickly become obsolete or unreliable if not regularly updated and monitored. This section delves into the various challenges that can hinder the reliability of AI systems, from data-quality issues to the elusive phenomenon of model drift. By understanding these challenges, we can better equip ourselves to address them and build AI systems that stand the test of time.

Data Quality

In the realm of AI and machine learning, the adage "garbage in, garbage out" holds particularly true. The quality of the data fed into a model determines the quality of its outputs. Poor data quality can manifest in various ways, each with its unique challenges and implications for model reliability.

Following are some of the most common data-quality issues impacting the reliability of AI models:

- **Incomplete Data:** Datasets often have missing values or incomplete records. This can be due to various reasons, such as sensor malfunctions, data-entry errors, or data-collection issues. When a model is trained on incomplete data, it may develop an incomplete or skewed understanding of the underlying patterns, leading to unreliable predictions.

- **Inaccurate Data:** Erroneous entries or mislabeled data points can actively mislead the model during training. For instance, in a dataset for image classification, if several images of cats are labeled as dogs, the model might struggle to differentiate between the two accurately.

- **Outdated Data:** The world is dynamic, and data that was relevant a few years ago might not reflect current realities. Using outdated data for training can result in a model that's out of touch with current trends, behaviors, or patterns.

- **Bias in Data:** Bias is a pervasive issue in AI. As we have seen in previous chapters, if the training data is not representative of the broader population or contains inherent biases, the model will likely inherit these biases. This can lead to unfair, skewed, or discriminatory predictions.

Here's how these can be addressed:

- **Data Audits:** Regularly auditing the dataset can help identify inconsistencies, missing values, or anomalies. Tools like pandas' `describe()` method in Python can provide a quick overview of the data's statistical properties, highlighting potential issues.

- **Data Imputation:** Techniques like mean imputation, median imputation, or even more advanced methods like k-nearest neighbors can be used to fill in missing values, ensuring the dataset is complete.

- **Outlier Detection:** Algorithms like the isolation forest or DBSCAN can detect and handle outliers, ensuring they don't skew the model's training.

- **Bias Detection and Mitigation:** Tools like Fairness Indicators or AI Fairness 360 can help in detecting and mitigating biases in datasets. We have already looked at these in detail in previous chapters.

Here's a simple example illustrating some techniques for ensuring data quality:

```
[In]:
import pandas as pd
from sklearn.impute import SimpleImputer
from sklearn.ensemble import IsolationForest

# Load data
data = pd.read_csv('data.csv')

# Data audit to get a quick overview
print(data.describe())

# Handling missing data using mean imputation
imputer = SimpleImputer(strategy='mean')
data_imputed = imputer.fit_transform(data)

# Detecting and removing outliers using Isolation Forest
iso_forest = IsolationForest(contamination=0.05)
outliers = iso_forest.fit_predict(data_imputed)
data_cleaned = data_imputed[outliers != -1]
```

In the preceding example, we first audit the data to understand its statistical properties. We then handle missing values using mean imputation and detect and remove outliers using the isolation forest algorithm. These steps are just a starting point, and depending on the dataset's specifics, one might need to employ more advanced techniques or tools.

Model Drift

Model drift, often referred to as "concept drift," occurs when the statistical properties of the target variable, which the model is trying to predict, change over time. This drift can lead to a decrease in model accuracy and reliability. It's a common phenomenon in dynamic environments where data patterns evolve. For instance, customer preferences in e-commerce can change seasonally, or financial transaction patterns can shift due to economic events.

There are several types of model drift, as follows:

- **Sudden Drift:** This happens abruptly, where the model's performance drops sharply. An example could be a sudden change in user behavior due to an external event, like a pandemic.

- **Incremental Drift:** Here, the change happens gradually over time. An aging machine on a production line producing slightly different outputs over years is an example.

- **Gradual Drift:** This involves alternating between two or more states. For instance, a website's user behavior might oscillate between two patterns during weekdays and weekends.

Detecting and addressing model drift is crucial for maintaining the reliability of AI systems. Regularly monitoring model predictions against actual outcomes can help in early detection. If the model's performance has degraded a lot, it might indicate model drift. Once identified, retraining the model with fresh data or adjusting its parameters can help realign it with the current data distribution. Here's a simple example of how to do something like this:

```
[In]:
from sklearn.metrics import accuracy_score

# Assuming 'model' is our trained model and 'X_test', 'y_test'
are test data and labels
predictions = model.predict(X_test)
initial_accuracy = accuracy_score(y_test, predictions)
```

```
# After some time, we test the model again
new_predictions = model.predict(X_new_test)
new_accuracy = accuracy_score(y_new_test, new_predictions)

# Check for significant drop in accuracy
if new_accuracy < initial_accuracy - threshold:
    # Retrain the model or take corrective action
    model.fit(X_new_train, y_new_train)
```

In this example, we monitor the model's accuracy over time. If there's a significant drop, we take corrective action by retraining the model. This is a simplified approach, and in real-world scenarios, more sophisticated drift detection mechanisms might be required.

Uncertainty in AI Models

AI models, especially those used in critical applications like health care, finance, or autonomous vehicles, are expected to make predictions with a high degree of confidence. However, in real-world scenarios, models often encounter situations or data points that were not present or were underrepresented in their training data. In such cases, the model's prediction might be accompanied by a significant level of uncertainty. This uncertainty can arise from various sources, such as the following:

- **Model Uncertainty:** Even with a perfect training dataset, the model might not capture all the underlying complexities of the data. This is especially true for simpler models or when the true data distribution is inherently complex.

- **Data Uncertainty:** Real-world data is often noisy, incomplete, or even erroneous. When models are trained on such data, their predictions can inherit this uncertainty.

- **Distributional Uncertainty:** This arises when the model encounters data that come from a different distribution than its training data. For instance, a model trained on images taken during the day might be uncertain when predicting on images taken at night.

Given the potential risks associated with acting on uncertain predictions in critical applications, it's imperative to quantify and communicate this uncertainty.

Uncertainty quantification (UQ) is a discipline that focuses on quantifying, characterizing, and managing uncertainties in computational and real-world systems. In the context of AI, UQ provides techniques to estimate the uncertainty associated with model predictions. Here's how UQ can be beneficial:

- **Confidence Intervals:** For regression tasks, UQ can provide confidence intervals around predictions, giving a range in which the true value is likely to lie.

- **Probabilistic Outputs:** For classification tasks, instead of providing a single class label, models can be trained to output a probability distribution over all possible labels, reflecting the model's confidence in its prediction.

- **Bayesian Neural Networks:** These are neural networks that are trained to output distributions rather than point estimates. They inherently capture the model's uncertainty about its predictions.

- **Model Ensembling:** Using an ensemble of models and observing the variance in their predictions can provide an estimate of uncertainty. If all models in the ensemble agree, the uncertainty is low. However, if they widely disagree, the uncertainty is high.

By incorporating UQ techniques, AI practitioners can ensure that their models not only make accurate predictions but also communicate the confidence level of these predictions. This added layer of transparency can be crucial in decision-making processes, especially in critical applications where the stakes are high.

Conclusion

In the journey through this chapter, we've delved deep into the intricacies of ensuring that AI models are both robust and reliable. The challenges posed by adversarial attacks, model overfitting, and data noise underscore the vulnerabilities inherent in AI systems. On the flip side, the uncertainties introduced by data-quality issues, model drift, and the very nature of real-world data highlight the complexities of ensuring model reliability. However, as we've seen, these challenges are not insurmountable. Through a combination of techniques ranging from data augmentation and regularization to continuous monitoring and uncertainty quantification, we can fortify our AI systems against these pitfalls.

The overarching message is clear: While AI offers transformative potential across various domains, its true power is harnessed only when models are both robust to adversarial perturbations and reliable in their predictions. As AI continues to permeate every facet of our lives, the principles and practices outlined in this chapter will be instrumental in building systems that not only perform well but also earn the trust of their users. As practitioners, it's our responsibility to ensure that the AI systems we deploy are as resilient as they are revolutionary.

CHAPTER 6

Conclusion

In this concluding chapter, we reach the culmination of our journey
through the realm of responsible AI, encapsulating the essence of our
exploration in three pivotal sections. First, we will synthesize the key
findings that have emerged from our thorough investigation, providing a
succinct overview of the critical insights and lessons that have unfolded
throughout this book. Next, we will issue a compelling call to action,
addressing developers, businesses, and policymakers, as we collectively
consider the responsibilities and opportunities that lie ahead in the
Responsible AI landscape. Lastly, we will share our final thoughts,
reflecting on the broader implications and significance of the Responsible
AI movement, inviting readers to ponder the path forward and the role
each of us can play in shaping a more ethical and equitable
AI-driven future.

Summary of Key Findings

In this comprehensive exploration of Responsible AI, we have journeyed
through the foundational chapters that lay the groundwork for building
ethical and accountable artificial intelligence (AI) systems. We began
with an insightful introduction that highlighted AI's vast potential and the
paramount importance of Responsible AI in today's world. Emphasizing
the core ethical principles guiding our path forward, we set the stage for a
profound examination of key issues.

Chapter 2 delved into the intricate realm of bias and fairness within AI systems. Our exploration took us through the complexities of recognizing and mitigating bias, offering a valuable toolkit for building equitable AI solutions. Chapter 3 illuminated the critical need for transparency and explainability in AI, providing both the conceptual framework and practical tools necessary to demystify the decision-making processes of AI models. Furthermore, this chapter underscored the importance of acknowledging and overcoming the challenges inherent in achieving these goals.

Chapter 4 ventured into the realms of privacy and security, unveiling the lurking concerns and vulnerabilities that AI systems introduce. By meticulously examining strategies to protect data, models, and overall system security, this chapter equipped us with the knowledge and tools required to navigate this intricate landscape. In the penultimate chapter, Chapter 5 underscored the importance of robustness and reliability within the realm of AI applications. We explored the methods for crafting durable models and approaches for conducting thorough assessments and validations of AI systems, thereby guaranteeing their adherence to the most stringent criteria of dependability. With each chapter, we've charted a course toward a future where AI serves as a responsible and ethical tool capable of advancing society while preserving our values and principles.

Role of Responsible AI in Business Adoption

Responsible AI can play a significant role in facilitating higher business adoption of AI technologies by addressing key concerns and barriers that often hinder adoption. Here's how responsible AI can help businesses adopt AI more effectively:

- **Ethical Assurance:** Responsible AI practices prioritize ethical considerations in AI development and deployment. This helps alleviate concerns related to the ethical implications of AI, such as bias, discrimination, and privacy violations. By ensuring that AI systems align with ethical standards, businesses can build trust with stakeholders and the public, making it easier to adopt AI solutions.

- **Legal and Regulatory Compliance:** Responsible AI helps businesses navigate the complex and evolving landscape of AI regulations and standards. It ensures that companies comply with existing and emerging laws related to AI, reducing the legal risks associated with non-compliance. Compliance with regulatory requirements instills confidence among decision-makers, encouraging them to embrace AI technologies.

- **Transparency and Accountability:** Responsible AI emphasizes transparency in AI decision-making processes. This means that businesses can explain how AI algorithms arrive at their conclusions, making AI more understandable and accountable. When stakeholders can scrutinize AI systems and their outputs, they are more likely to accept and adopt these technologies.

- **Risk Mitigation:** By addressing risks associated with AI adoption, such as algorithmic bias, security vulnerabilities, and unintended consequences, responsible AI helps mitigate these risks. This risk mitigation not only protects the business but also ensures that AI is deployed more reliably, which can encourage broader adoption.

- **Customer Trust:** Businesses that prioritize responsible AI gain the trust of their customers. When consumers believe that AI is used responsibly, they are more likely to engage with AI-powered products and services. Higher levels of trust can lead to increased adoption rates as customers feel more comfortable using AI-driven solutions.

- **Employee Buy-In:** Employees are often directly affected by AI adoption in the workplace. Responsible AI practices consider the impact on employees, addressing concerns related to job displacement and ensuring that AI is used to augment human capabilities. Engaged and satisfied employees are more likely to support AI adoption efforts within the organization.

- **Reduced Bias and Discrimination:** One of the central concerns with AI is bias and discrimination in decision making. Responsible AI methods aim to reduce these biases, making AI systems fairer and more equitable. By actively addressing bias concerns, businesses can encourage adoption among diverse user groups and avoid negative publicity.

- **Strategic Advantage:** Adopting responsible AI can be a strategic advantage. Businesses that prioritize ethical and responsible AI practices often differentiate themselves in the market, attract socially conscious customers, and align with investor preferences. This strategic positioning can lead to increased market share and revenue.

- **Long-Term Viability:** Businesses that embrace responsible AI are more likely to ensure the long-term viability of their AI initiatives. This is because they are better equipped to handle ethical and regulatory challenges, reducing the risk of costly setbacks or discontinuation of AI projects.

However, responsible AI, when not implemented properly, can potentially decrease business adoption for several reasons:

- **Reputation Damage:** Irresponsible AI practices can lead to public backlash and damage a company's reputation. If customers perceive that a business is using AI in ways that are unethical, biased, or harmful, they may boycott the company's products or services, leading to a loss of market share and revenue.

- **Legal and Regulatory Risks:** Failing to implement responsible AI can result in legal and regulatory risks. Many jurisdictions are introducing laws and regulations to govern AI, and non-compliance can lead to significant fines and legal troubles for businesses. This can deter companies from adopting AI technologies.

- **Loss of Customer Trust:** Trust is crucial in business, and using AI irresponsibly can erode customer trust. Customers want to know that their data is being handled ethically and that they expect transparency in AI-driven decision making. If a company doesn't meet these expectations, customers may take their business elsewhere.

- **Bias and Discrimination:** If AI systems are not properly designed and tested for bias, they can reinforce and perpetuate existing societal biases and discrimination. This can lead to discrimination lawsuits and public condemnation, making businesses hesitant to adopt AI.

- **Security Vulnerabilities:** Poorly implemented AI systems can introduce security vulnerabilities. If AI systems are not adequately secured, they can become targets for hackers, potentially leading to data breaches and other security incidents that harm a company's reputation and bottom line.

- **Inefficient or Ineffective Operations:** If AI is implemented without considering its ethical and societal implications, it can lead to inefficient or ineffective operations. For example, automated decision making that doesn't take into account ethical considerations may lead to suboptimal outcomes, impacting the business's efficiency and competitiveness.

- **Employee Resistance:** Employees may resist the adoption of AI technologies if they perceive that these technologies are being used irresponsibly or if they fear that AI will replace their jobs without due consideration for retraining or redeployment.

- **Investor Concerns:** Responsible AI practices are increasingly a concern for investors and shareholders. Companies that do not take responsible AI seriously may face pressure from investors, which can negatively impact stock prices and the overall valuation of the business.

- **Market Barriers:** Some markets and industries have stringent ethical and regulatory requirements for AI adoption. Businesses that fail to meet these requirements may find it challenging to enter or expand in such markets.

- **Stifling Innovation:** Focusing solely on profit-driven AI adoption without responsible considerations can lead to short-term gains but stifle long-term innovation. Ethical and responsible AI practices can foster innovation that aligns with societal values and customer expectations.

Call to Action for Developers, Businesses, and Policymakers

Implementing responsible AI is a complex endeavor that necessitates the combined efforts of developers, businesses, and policymakers. It is a call to action that aims to foster the ethical and beneficial use of artificial intelligence. In this comprehensive exploration, we will delve into the specific responsibilities and actions that each of these stakeholders must undertake to create a responsible AI ecosystem.

Developers

Developers play a crucial role in implementing Responsible AI systems. Here are some key aspects of the role developers play in ensuring responsible AI implementation:

- **Embed Ethics from the Start:** The journey toward responsible AI begins with developers. They must prioritize ethical considerations from the very inception of AI development projects. During the design and coding phases, developers should contemplate the potential biases, fairness, and consequences of the AI algorithms they are creating. By doing so, they can preclude the emergence of ethical issues that may arise later in the development process.

- **Continuous Education:** The field of AI ethics is dynamic and ever-evolving. Developers must commit to ongoing education to stay abreast of the latest developments in responsible AI. This includes participating in training programs and workshops that emphasize ethics, fairness, and transparency in AI. Continuous education ensures that developers are equipped with the knowledge and skills needed to navigate the ethical complexities of AI.

- **Audit and Test:** Responsible AI demands vigilance. Developers should establish a regular audit and testing process for AI systems to identify and rectify any ethical shortcomings. This includes assessing the AI for bias, fairness, and transparency issues and taking corrective actions as necessary. Rigorous testing procedures are essential to maintain AI systems that meet the highest ethical standards.

- **Explainability and Transparency:** Developers should actively work to enhance the transparency and explainability of AI systems. Users should be able to understand how and why AI algorithms make decisions. Achieving transparency not only builds trust but also empowers users to make informed choices when interacting with AI-powered applications.

- **User Consent and Control:** Respect for user privacy and autonomy is paramount. Developers should prioritize mechanisms that give users the ability to provide informed consent and exercise control over how their data is used by AI systems. This entails designing user-friendly interfaces that facilitate data management and privacy settings.

Businesses

Businesses play a significant role in implementing Responsible AI practices. Responsible AI implementation is not solely the responsibility of developers or data scientists; it requires a holistic approach involving various stakeholders within an organization. Here are several ways businesses can contribute to the responsible implementation of AI:

- **Corporate Responsibility:** Ethical AI isn't merely a legal requirement; it's a corporate responsibility. Businesses must recognize this responsibility and establish AI governance frameworks that prioritize ethical considerations throughout the AI development lifecycle. This commitment starts at the executive level and permeates throughout the organization.

- **Diverse Teams:** Encourage diversity within AI development teams. Diverse teams bring varied perspectives and experiences, which can help recognize and address potential biases and fairness concerns in AI systems. Inclusive teams foster a culture of ethical awareness and innovation.

- **Transparency in Products:** Businesses must be transparent about the capabilities and limitations of their AI products and services. Avoid making exaggerated claims about AI capabilities that could mislead users. Honesty and transparency are essential for building and maintaining trust.

- **Ethical Data Use:** Establish clear policies governing data collection, usage, and sharing. Ensure that data is used responsibly and in compliance with privacy regulations. Businesses should be accountable for the ethical handling of user data and should be transparent about their data practices.

- **Collaboration:** Collaboration is key to advancing responsible AI. Businesses should collaborate with other organizations and stakeholders to share best practices and develop industry standards for responsible AI. Open dialogue and knowledge sharing can accelerate the adoption of ethical AI practices across industries.

Policymakers

Policy makers play a crucial role in shaping the ethical and legal frameworks that govern the development, deployment, and use of artificial intelligence (AI) technologies. Their decisions have a significant impact on how AI is developed and employed in various sectors. Here are several key roles policy makers play in implementing responsible AI:

- **Regulation and Enforcement:** Policymakers play a pivotal role in shaping the responsible AI landscape. They should develop and enforce regulations that govern the ethical use of AI. These regulations should outline the responsibilities of AI developers and businesses and set clear expectations for ethical behavior.

- **Transparency Requirements:** Mandate transparency and explainability in AI systems. Organizations should be required to disclose how their AI algorithms make decisions, providing users with insights into the decision-making process. Transparency requirements promote accountability.

- **Accountability:** Policymakers should establish mechanisms for holding organizations accountable for unethical AI practices. This includes the implementation of fines and penalties for violations of AI ethics regulations. Accountability measures serve as deterrents against unethical behavior

- **Data Privacy Laws:** Strengthen data privacy laws to protect individuals' rights and ensure that AI systems do not infringe upon privacy. Policymakers should continually update and adapt privacy regulations to keep pace with AI advancements.

- **Public Awareness:** Educate the public about AI and its implications. Foster a sense of responsibility and ethical awareness among AI users. Policymakers can play a role in facilitating public discourse on AI ethics and promoting responsible AI use.

- **Support Research:** Allocate resources to support research on responsible AI, including bias detection, fairness, and transparency. Policymakers can fund research initiatives and collaborate with academic and industry experts to advance our understanding of ethical AI practices.

In conclusion, implementing responsible AI is a shared responsibility that necessitates the commitment and concerted efforts of developers, businesses, and policymakers. By embedding ethical principles into AI development, promoting transparency, and enforcing regulations, we can harness the potential of AI while minimizing its risks and ensuring that it benefits society as a whole. Together, we can build a future where AI serves as a force for good, upholding our values and ethical standards.

Final Thoughts

Future Outlook

The future outlook of responsible AI is promising and likely to continue evolving in several key directions:

- **Regulation and Compliance:** Governments and regulatory bodies around the world are expected to introduce more-stringent regulations related to AI ethics and responsible AI practices. This will likely require businesses to adhere to specific standards and guidelines for the development and deployment of AI systems.

- **Ethical AI Education:** As AI becomes more integrated into society, there will be a growing need for education and training in AI ethics and responsible AI practices. Educational institutions and organizations will offer courses and programs to equip individuals and professionals with the knowledge and skills needed to work with AI in an ethical and responsible manner.

- **Auditing and Certification:** The development of auditing and certification processes for AI systems will likely become more common. Independent organizations may emerge to evaluate AI systems for fairness, transparency, and other ethical considerations, similar to how certifications like ISO standards operate.

- **Human–AI Collaboration:** The future will see increased collaboration between humans and AI. Responsible AI will involve designing AI systems that augment human capabilities and work in tandem with humans, rather than replacing them entirely.

As we wrap up our journey through this introduction to responsible AI, it's crucial to reflect on the profound impact this field has on our world. We've embarked on a quest to understand the promises and perils of artificial intelligence and the pivotal role that responsible AI plays in shaping its trajectory.

As we move forward, let's remember that the responsible development and deployment of AI is a collective endeavor. It requires collaboration among researchers, engineers, policymakers, ethicists, and society as a whole. Together, we can ensure that AI is a force for good, enhancing our lives, respecting our values, and contributing to a brighter and more equitable future. So, as you embark on this journey into the world of responsible AI, we encourage you to approach it with curiosity, responsibility, and a commitment to making AI work for the benefit of all.

Index

A

Achieving reliability, challenges
    data quality, 152–154
    model drift, 154–156
    uncertainty, 156–158
Achieving robustness, challenges
    adversarial examples,
        transferability, 148–151
    model overfitting, 141–144
    outliers and noise, 145–148
    sensitivity, input
        variations, 139–141
Adversarial attacks, 20, 120,
    122–125, 137, 139, 151
Adversarial Debiasing
    algorithm, 52
Adversarial examples
    addressing ways
        adversarial training, 149
        input preprocessing, 149
        model diversity, 150
        randomized defenses, 149
    "black-box" attacks, 148
    implications
        defense challenges, 149
        model ensemble
        vulnerability, 149
        wider attack surface, 148
    transferability, 148
Adversarial robustness, 137
Adversarial testing, 37
Adversarial training, 11, 20, 122,
    140, 149
AI-driven algorithms, 136
AI Fairness 360 toolkit, 52, 54
AI security risks mitigation
    backdoor detection and
        removal, 129, 130
    conclusion, 131
    defense mechanisms,
        adversarial training
        feature squeezing, 124
        gradient masking, 123
        TensorFlow, 122
    input filtering, evasion attacks,
        127, 128
    model hardening, 125, 127
    monitoring and auditing,
        130, 131
    multifaceted, 122
Algorithmic bias, 31, 161
Amazon's gender-biased hiring
    tool, 32
Artificial intelligence (AI), 159
    accuracy and performance, 133
    AI-driven algorithms, 136

Printed in the United States
by Baker & Taylor Publisher Services